Lecture Notes
in Economics and
Mathematical Systems

Managing Editors: M. Beckmann and W. Krelle

236

Giancarlo Gandolfo
Pietro Carlo Padoan

A Disequilibrium Model of Real and Financial Accumulation in an Open Economy

Theory, Evidence, and Policy Simulations

Springer-Verlag
Berlin Heidelberg New York Tokyo 1984

Editorial Board

H. Albach M. Beckmann (Managing Editor) P. Dhrymes
G. Fandel J. Green W. Hildenbrand W. Krelle (Managing Editor) H.P. Künzi
G.L. Nemhauser K. Ritter R. Sato U. Schittko P. Schönfeld R. Selten

Managing Editors

Prof. Dr. M. Beckmann
Brown University
Providence, RI 02912, USA

Prof. Dr. W. Krelle
Institut für Gesellschafts- und Wirtschaftswissenschaften
der Universität Bonn
Adenauerallee 24–42, D-5300 Bonn, FRG

Authors

Prof. Giancarlo Gandolfo
Prof. Pietro Carlo Padoan
Faculty of Economics, University of Rome
Via del Castro Laurenziano 9, I-00161 Rome, Italy

ISBN 3-540-13889-7 Springer-Verlag Berlin Heidelberg New York Tokyo
ISBN 0-387-13889-7 Springer-Verlag New York Heidelberg Berlin Tokyo

This work is subject to copyright. All rights are reserved, whether the whole or part of the material is concerned, specifically those of translation, reprinting, re-use of illustrations, broadcasting, reproduction by photocopying machine or similar means, and storage in data banks. Under § 54 of the German Copyright Law where copies are made for other than private use, a fee is payable to "Verwertungsgesellschaft Wort", Munich.

© by Springer-Verlag Berlin Heidelberg 1984
Printed in Germany

Printing and binding: Beltz Offsetdruck, Hemsbach/Bergstr.
2142/3140-543210

Preface

This is the fourth version of a model that five years ago we set out to build and estimate along the lines of the continuous time approach clarified in chapter 1. Previous versions appeared in journal articles and conference proceedings, where the space is notoriously limited. Therefore we welcome the possibility of publishing a book-length treatment of this fourth version, so that we can describe its theoretical and empirical aspects in some detail. Although we have worked closely together and accept joint responsibility for the whole book, chs. 1 and 2 and appendix I have been written by G. Gandolfo, whilst chs. 3 and 4 and appendix II have been written by P.C. Padoan.

Different parts of this version of the model have been discussed in various lectures at the European University Institute (Florence) in 1984, in a seminar organized by the Bank of Italy (Sadiba, Perugia, Italy, February 16-18, 1984), in the second Viennese Workshop on Economic Applications of Control Theory (Vienna, May 16-18, 1984), and in the sixth annual Conference of the Society for Economic Dynamics and Control (Nice, France, June 13-15, 1984). In all of these we received helpful comments; similarly helpful were the comments of Clifford R. Wymer, who, however, is absolved of any responsibility.

Grants by the Consiglio Nazionale delle Ricerche, the Ministry of Public Instruction, and the University of Rome at different times gave us the indispensable financial support for bearing the expenses of our research. They are gratefully acknowledged, but of course none of these institutions is responsible for the views expressed here.

Miss Anna Maria Olivari, in addition to carrying out all the secretarial work, typed endless drafts and prepared the final camera-ready typescript with flawless grace: if any misprints remain, it is our fault, not hers.

University of Rome
Faculty of Economics and Commerce

CONTENTS

PREFACE

Chapter 1

METHODOLOGICAL PROBLEMS   1

   1.1 Why "Small" Models   1
   1.2 Why Continuous Models   4
   1.3 Plan of the Book   11

Chapter 2

THE MODEL   13

   2.1 An Overwiew   13
   2.2 The Single Equations   14
      2.2.1 The Real Sector   19
      2.2.2 The Price and Wage Sector   24
      2.2.3 The Financial Sector   26
      2.2.4 Policy Reaction Functions   29
      2.2.5 Final Considerations   34
   2.3 Steady-state, Comparative Dynamics, Stability, and Sensitivity   36

Chapter 3

EMPIRICAL RESULTS   42

   3.1 Parameter Estimates   42
      3.1.1 Adjustment Parameters   43
      3.1.2 Elasticities and Other Parameters   51
   3.2 Stability and Sensitivity   62
   3.3 Predictive Performance   71

Chapter 4

POLICY SIMULATIONS   77

   4.1 Introduction   77
   4.2 Antiinflationary Policies   78
   4.3 Endogenous Exchange Rate Determination   101
   4.4 Exogenous Variables Behaviour Altered   108
   4.5 Global Strategies   113

APPENDIX I   126

   I.1 Derivation of the Steady-state and Comparative Dynamics Results   126
   I.2 Linearization about the Steady-state   133
   I.3 On the Elimination of $\tilde{Y}$ 142

APPENDIX II 144

   II.1 Data Sources and Definitions 144
      II.1.1 Sources of Data 144
      II.1.2 Definition of Series 144
   II.2 On the Endogenous Estimation of Productivity 147
   II.3 Simulation Procedures 148
   II.4 Whither Simulations? 153

REFERENCES 158

AUTHOR INDEX 167

SUBJECT INDEX 169

CHAPTER 1

Methodological Problems

1.1. Why "Small" Models

Everyone knows that in econometric model building for a long time now there has been a tendency towards gigantism, in which one of the recommedantions for a model was its size (measured by the number of equations). This tendency still continues, but the careful observer will now have realized that another idea has been gaining ground, namely that macrodynamic medium-term econometric models, small in size (from one to a few dozen equations) and embodying considerably more economic theory than is usually possible, are complementary, or even preferable, to large models[1] for the purposes of policy analysis and simulations: see, for example, Bergstrom and Wymer, 1976; Wymer, 1976, 1979; Reserve Bank of Australia,1977;Committee on Policy Optimisation, 1978; Deleau et al., 1980, 1982; Masson, Rose and Selody, 1980; Richard, 1980; Gandolfo, 1981; Bergstrom, 1982; Kirkpatrick, 1983; Gandolfo and Padoan, 1982, 1983; and the references therein.

It is interesting to note that this idea is not only circulating amongst academics, but has also been put to work by central banks and international organisations. For example,the Reserve Bank of Australia has developed and is using such a model, now in its second edition (RBA, 1977; Jonson and Trevor, 1981; Taylor, 1982); the Bank of Canada is currently building another (Masson, Rose and Selody, 1980; Rose, Selody and Masson, 1982); the IMF has built several such models for separate countries (the best being that of Knight and Wymer, 1978).

A related problem concerns the discrepancy between theoretical and econometric modelling. Many economists — ourselves among them — feel

---

[1] A review of the small vs large models controversy is contained, for example, in Gandolfo (1984a).

that, at macroeconomic level, such a discrepancy exists. Theoretical modelling proceeds by means of sophisticated qualitative analyses of very small models (normally not exceeding two final dynamic equations, to enable the theoretician to use phase-plane techniques), with all the advantages, but also the limitations, caused by the oversimplifications made necessary by this smallness: although these models may give important theoretical insights, the oversimplification inherent in their smallness does not commend them for practical applications, such as are required in policy analysis. Econometric modelling proceeds by means of numerical analyses of huge models, with all the advantages but also the limitations inherent in this approach (overparameterization, loss of intellectual control, possible lack of a unifying theoretical structure, piecemeal modelling etc.: see Gandolfo, 1984a).

In our opinion, small econometric models may help to eliminate this discrepancy, on the one hand by maintaining all the advantages of a well-founded and consistent theoretical structure and amenability to qualitative analysis, and on the other by allowing their use for policy analysis and simulations once a satisfactory estimation of their parameters has been obtained.

The ideas under consideration seem to point to small-size models (from one to a few dozen equations, as we said) with the following characteristics:

i) They must have a sound and consistent theoretical structure. Economic theory implies constraints on functional form and cross-equation restrictions on coefficients, and these should be imposed on the model. This in turn implies that

ii) the models under consideration ought to be estimated econometrically by using system (i.e., full information) methods. Among these methods, full-information maximum likelihood (FIML) is preferable to the other major full-information technique (three-stage least squares). This is because it has the advantage of using in the estimation procedure a wide range of *a priori* information, pertaining not only to each equation individually, but also to several equations simultaneously, such as non-linear within and across equation restrictions on the parameters. Furthermore, a lot is known about the properties of FIML estimators. It goes without saying that the model will have to be specified with great care in view of the well-known sensitivity of full information methods to specification errors. But, contrary to a widespread opinion, we do not view this sensitivity as a drawback, because it compels the model builder to specify and evaluate the model as a unified whole rather than as a series of equations. It also helps him to detect possible specification errors, which might remain undetected when using other estimation methods. In addition, the model as a whole *is* sensitive to a change in any one equation, and estimation with single equation methods conceals this fact. The "robustness" of these estimates, in particular of OLS estimates, diverts attention from the fact that estimation results are anything but robust to changes in specification.

iii) The models should contain an adequate consideration of stock-flow interaction as well as of interaction between real and financial variables. This means a theoretical eclecticism which does not embrace any one of the competing views but tries to take the best of each of them. In the real world there are not only stock adjustments or pure flows, but both; nor is it true that only financial variables or only real variables matter: both do.

iv) The models should embody an adequate consideration of dynamic processes of adjustment to disequilibrium. Although the notion of equilibrium is still central to the way of thinking in the profession[2] the real world is practically never in equilibrium even if, hopefully, it is moving towards an equilibrium (which in turn may be a *moving equilibrium*, possibly a steady-state growth path). The models should reflect this fact and so be conceived from the point of view of *disequilibrium dynamics*.

v) The models should possess satisfactory medium-long term properties. An implausibile long-run behaviour (for example, explosive oscillations) could indicate a structural defect, and seriously impair

---

[2] For some considerations on this point, see Gandolfo (1981, pp. 19-20).

the usefulness of the model for policy analysis and simulation. Since the second world war, most countries have followed a growing path with more or less pronounced oscillations. Therefore, the models under consideration should be capable of generating a *stable cyclical growth path*.

1.2 Why Continuous Models

We now come to the gist of the matter and begin by observing that the trend towards small models and the trend towards *continuous time* models are closely correlated. Perhaps this is because continuous models are better suited to satisfy the requirements described in points i)-v) above. It is interesting to note that, quite independently of each other, the first modern writers to advocate the use of continuous models were an economic theorist (Goodwin, 1948, pp. 113-4) and an econometrician (Koopmans, 1950). It is also interesting to note that Koopmans put forward the idea of formulating econometric models in continuous time in a short essay contained in the volume that laid the foundation for modern econometric methodology. In this essay Koopmans also illustrated some of the adavantages of these models over those formulated in discrete time. These advantages led Marschak to state, in the general introduction to the volume mentioned, that "if proper mathematical treatment of stochastic models can be developed, such models [that is, those formulated in continuous time] promise to be a more accurate and more flexible tool for inference in economics than the discrete models used heretofore" (Marschak, 1950, p. 39; square brackets are ours). However — except in occasional works — these suggestions were not followed up and the topic was taken up again and further developed only in the seventies.

The choice of the kind of "time" (continuous or discrete[3]) to be

---

[3] What follows draws heavily on Gandolfo (1981, ch. 1), with some additional considerations concerning points 5 and 8 below.

used in the construction of dynamic models is a moot question (see, for example, Gandolfo, 1980, *passim*). We must remember that such a choice implies the use of different analytical tools: differential equations in the continuous case, difference equations in the discrete one. This choice is not unbiased with respect to the results of the model, given the different behaviour of the solutions of these types of functional equations and the different nature of the stability conditions.

In our opinion, mixed differential-difference equations are much more suitable for an adequate treatment of dynamic economic phenomena than differential equations alone, or difference equations alone. However, the formal difficulty of mixed equations greatly limits their use in theoretical studies, and, for the moment, the impossibility of obtaining econometric estimates of their parameters excludes their use in applied research. Therefore a choice has to be made between the two basic formulations, that in terms of differential equations and that in terms of difference equations. But until now such a choice could be made only at the level of theoretical studies, and not at the level of applied research, because current econometric techniques for the estimation of complete models are based on discrete time; therefore, in order to estimate a model, it was necessary to formulate it in discrete time. If it had been conceived in continuous time, then one had to approximate it with a discrete model (and approximations which have been used in the past are not acceptable once the problem is formulated rigorously). The consequence is a discrepancy between theoretical modelling and econometric modelling, which is definitely not favourable to the advancement of scientific knowledge.

Fortunately, the problem of statistical inference in continuous-time dynamic models has now been dealt with satisfactorily (Bergstrom (ed.), 1976; Bergstrom, 1983, 1984; Gandolfo, 1981; Wymer, 1972, 1976)[4] thus allowing a rigorous estimation of the parameters of sys-

---

[4]Alternative approaches to that used by these authors have been suggested: see,

tems of stochastic differential equations on the basis of samples of discrete observations such as are available in reality. The new methodology marks a great advance, because it allows the economist who has chosen a continuous model at the theoretical level, to estimate its parameters rigorously and independently of the observation interval. This means that the bias in favour of discrete models (imposed by current econometric methods and not by economic theory) disappears. This leaves the researcher free to formulate his model as he thinks best and to estimate its parameters whether this formulation is discrete or continuous.

We will now expound the main arguments in favour of the use of continuous models, which, for semplicity, we have grouped into eight categories.

1) Although individual economic decisions are generally made at discrete time intervals, it is difficult to believe that they are coordinated in such a way as to be perfectly synchronized (that is, made at the same moment and with reference to the same time interval as postulated by period analysis). On the contrary, it is plausible to think that they overlap in time in some stochastic manner. As the variables that are usually considered and observed by the economist are the outcome of a great number of decisions taken by different operators at different points of time, it seems natural to treat economic phenomena as if they were continuous.

2) A specification in continuous time is particularly useful for the formulation of dynamic adjustment processes based on excess demand, a plausible discrete equivalent of which is often difficult to find.

As an example (Wymer, 1976) let us consider the case where the interest rate changes as a function of the excess demand for real bal-

---

for example, Bailey, Hall and Phillips (1980) on the one hand and Harvey and Stock (1983) on the other. However, the relative merits of the various approaches have not yet been assessed, and the approach followed in the present book is that generally used in the actual estimation of continuous time models.

ances:

$$Dr = \alpha(m - \frac{M}{P}),$$

or, in logarithmic terms

$$D\log r = \alpha(\log \hat{m} - \log \frac{M}{P}) = \alpha \log(\frac{\hat{m}}{M/P}), \tag{a}$$

where D denotes the operator $\frac{d}{dt}$, $\hat{m}$ is the demand for real money balances (a function of income and of the interest rate, $\hat{m} = a\, Y\, r^\beta$), and the other symbols have the usual meanings. For estimation purposes the following discrete approximation is usually adopted

$$\Delta \log r_t = \alpha[\log \hat{m}_t - \log(\frac{M}{P})_{t-1}] = \alpha(\log Y_t + \beta \log r_t + \log a - \log M_{t-1} + \log P_{t-1}). \tag{b}$$

However, eq. (b) is not a satisfactory representation of eq. (a), because the demand function contains a *flow* variable (income) which is measured over the observation interval. The demand function also contains "point" or "instantaneous" variables, as the interest rate (which refers to the end of the current period) and real money balances (which are a stock variable and refer to the end of the previous period). It can be shown that a better approximation to (a) is given by

$$\Delta \log r_t = \alpha[\log Y_t + \frac{1}{2}\beta(\log r_t + \log r_{t-1}) + \log a - \frac{1}{2}(\log M_t + \log M_{t-1})$$

$$+ \frac{1}{2}(\log P_t + \log P_{t-1})], \tag{c}$$

but it is also possible to obtain and estimate the exact discrete equivalent model to the continuous model.

3) What has been said at Point 2) is connected with another advantage of continuous models: the estimator of these models is independent of the observation interval. What's more, it explicitly takes into account the fact that a flow variable cannot be measured instantaneously, so that what we actually observe is the integral of such a variable over the observation period (this allows a correct treatment of stock-flow models). These properties do not hold for the discrete models usually employed, which therefore must be formulated explicitly in relation to the data which are available or which one wishes to use. So, for example, a model built to be estimated with quarterly data will be different from one built to be estimated with

annual data.

4) A further difficulty of discrete analysis is that usually there is no obvious time interval that can serve as a "natural" unit. Lacking this, the assumption of a certain fixed period length may unwittingly be the source of misleading conclusions. Thus, it is necessary to check that no essential result of a discrete model depends on the actual time length of the period (the model should give the same results when such a period is, say, doubled or halved). But, if the results are unvarying with respect to the period length, they should remain valid when this length tends to zero (that is when one goes from discrete to continuous analysis). Some authors attach a special importance to this property and argue that the test of the invariance of results with respect to the length of the period is fundamental in order to ascertain whether a discrete model is well defined and consistent. Moreover, the lack of clarification, in such models, of whether equilibrium obtains at the beginning or at the end of the period, may give rise to confusion between stock equilibria and flow equilibria.

5) The partial adjustment functions discussed under 2), above, may have very high adjustment speeds and therefore very short mean time-lags with respect to the observation period. Because of this, it may happen that, when the variables are measured in discrete time, the desired value practically coincides with the observed value over the period, so that it is not possible to obtain an estimate of α. On the contrary, with the continuous formulation, it is always possible to obtain asymptotically unbiased estimates of α even for relatively long observation periods.

This possibility has important implications, especially when adjustment speeds play a crucial role, for example to determine which markets clear more rapidly, or to give a precise empirical content to the approach of "synergetics".

As regards the first point, much theoretical debate is based on different *a priori* assumptions on the relative speeds of adjustment in goods and asset markets. The asset market approach to the balance of payments and the exchange rate, for example, is based on the assumptions of continuous asset market equilibrium and of perfect capital mobility (extreme version) or, in a less extreme version, on the assumption that the adjustment speeds of asset markets are much higher than the adjustment speeds of goods markets. These assumptions, when not imposed on purely *a priori* grounds in the theoretical models, are tested by using standard discrete methods, which rule out the possibility that adjustment may occur faster than the unit period inherent in the data (Horne, 1983). The issue is so important that the availability of rigorous estimates of adjustment speeds independently of the observation interval should be welcome.

As regards the second point, the approach of synergetics — which, though discussed mainly in physics, chemistry and biology (see Haken, 1978, 1983), seems to offer fruitful insights also to economists (see Gandolfo and Padoan, 1983a; Medio, 1983; Silverberg, 1983) — divides the variables of any system into rapid motion (or "slaved") and slow motion (or "order") variables. It then reasonably assumes ("adiabatic appoximation") that the former are always near equilibrium so that *the evolution of the system is determined by the latter*; in other words, the order variables "slave" the remaining ones ("slaving principle"). The obvious implication is that the slow-motion or "order" variables are, in some sense, *more fundamental* than the rapid-motion or "slaved" variables.

We would like to stress the importance of the fact that, in economics, the distinction between "slaved" and "order" variables can be rigorously made (at a given confidence level) by means of empirical estimates (independently of the time unit chosen) of the adjustment speeds instead of *a priori* assumptions only.

6) The use of a continuous model may allow a more satisfactory treatment of distributed lag processes. In a discrete model the disturbances in successive observations are usually assumed to be independent, but this assumption can be maintained only if the size of the time unit inherent in the model is not too small relative to the observation period. The assumption of independence thus entails a lower limit on the permissible size of the inherent time unit, and

this precludes the correct treatment of a number of economic problems. The lags in the system are not always integral multiples of one time unit whose size is compatible with the assumption of independence. As it may happen that distributed time-lags with a lower time limit of almost zero have to be considered, a continuous time specification is more correct.

7) From the analytical point of view, differential equation systems are usually more easily handled than difference systems.

8) The availability of a model formulated as a system of differential equations enables its user — once the parameter estimates have been obtained — to get forecasts and simulations for any time interval, and not only for the time unit inherent in the data. In fact, the solution of the system of differential equations yields continuous paths for the endogenous variables, given the initial conditions concerning these variables and the time paths of the exogenous variables[5].

In discrete models it is usual to make the distinction between "single-period" (or "static") and "multi-period" (or "dynamic") forecasts. The former are those obtained by letting the lagged endogenous variables take on their actual values; the latter are those obtained by letting the lagged endogenous variables take on the values forecast by the model for the previous period(s). It is a well-known fact that dynamic forecasts are generally less good than single-period forecasts because the errors cumulate.

The equivalent distinction in continuous models can be made according to whether the solution of the differential equation system is (a) recomputed each period or (b) computed once and for all. In case (a) the differential equation system is re-initialized and solved n times (if we want forecasts for n periods), each time by using the observed

---

[5] These paths must be continuous functions, so that — if we are dealing with *ex post* forecasts — the series of discrete observations will have to be interpolated (see Gandolfo, 1981, ch. 3, section 3.4.2 and Wymer, "APREDIC" program and Manual). If we have to produce *ex ante* forecasts, the presumed behaviour of the exogenous variables can easily be represented by analytic functions of time.

values of the endogenous variables in period t as initial values[6] in the solution, which is then employed to obtain forecasts for period $t+1$. This is equivalent to the single-period forecasts in discrete models. In case (b) the observed values of the endogenous variables for the starting period are used as initial values in the solution of the differential equation system, which is then employed for the whole forecast period. This is equivalent to the dynamic forecasts in discrete models.

It goes without saying that single-period forecasts can be used either for *ex post* forecasting (for example, to examine the in-sample predictive performance of the model) or to produce *ex ante* forecasts for one period ahead. But, if one wants to produce *ex ante* forecasts for more than one period ahead and/or for a time interval different from that inherent in the data, then dynamic forecasts must be used; these, of course, can also be used for *ex post* forecasting.

1.3 Plan of the Book

As regards our model, a few years ago we set out to build and estimate a continuous time macrodynamic model of the Italian economy[7]; its initial version and results were presented in Gandolfo (1981) and in Gandolfo and Padoan (1980, 1981a). A second version was presented in Gandolfo and Padoan (1981b, 1981c, 1982a, 1982b), and a third version in Gandolfo and Padoan (1983a, 1983b). This is the fourth version, preliminary results of which were presented in Gandolfo (1984b) and in Gandolfo and Padoan (1984a, 1984b); further improvements will be possible in the future.

The organization of the volume is the following: in the next chapter

---

[6] The initial value problem is slightly more complicated than that (see Gandolfo, 1981, ch. 3, section 3.4.2), but the exposition would not be significantly altered if these complications were taken into account. The same remark applies to "dynamic" forecasts below.

[7] Other continuous models of the Italian economy are Tullio's (1981; see also the Comment by Gandolfo and Padoan, 1982c and the Reply by Tullio) and Chiesa's (1979).

we first give an overview of the theoretical model, which is then described in detail in section 2.2, where we the develop it from the ground up[8]; in section 2.3 we examine the qualitative properties of the model, with particular reference to its steady-state. In chapter 3 we present and comment on the results of the estimates; the stability and sensitivity of the model and its predictive performance are also fully analyzed. In chapter 4 we present several policy simulation exercises for the purpose of obtaining further insights into the properties of the model. Appendix I contains a mathematical analysis of the qualitative properties of the model and other technical matters concerning the theoretical part; Appendix II contains the data sources and definitions, a general note on simulation exercises, the simulation procedures, and other details on the empirical part.

---

[8] It may interest the reader to know the main changes with respect to the previous versions, leaving aside several modifications in the specification of the equations. The financial sector is more developed, for it includes bank advances, interest rate, money demand and supply, capital movements, public sector's borrowing requirement, international reserves (all these variables are endogenously determined); the interrelationships between real and financial variables are better specified; government expenditure and taxation are explicitly considered in addition to the monetary authorities' reaction function, which has been radically modified.

CHAPTER 2

The Model

2.1 An Overview

The model considers stock-flow behaviour in an open economy in which both price and quantity adjustments take place. Stocks are introduced with reference to the real sector (where adjustments of fixed capital and inventories to their respective desired levels are present) and to the financial sector which includes the stock of money, the stock of commercial credit, the stock of net foreign assets and the stock of international reserves. Real and financial feedbacks are, therefore, largely considered in the model. Government expenditure and revenues (taxation) are also present so that the effects of endogenous public deficits are included.

Quantity behaviour equations are considered for the traditional macroeconomic variables in real terms (private consumption, net fixed investment, imports and exports of goods and services, inventories, net domestic product). Expectations are present through an adaptive mechanism concerning expected output.

A price block is included, which determines the domestic price level, the nominal wage rate, and the export price level. Endogenous determination of the last was considered crucial for an export-led economy such as Italy's, while wage-price spiral effects are explicitly taken into account. The specification of the financial sector was completed by the inclusion of an interest-rate determination equation.

Although the model is a closely interlocked system of simultaneous differential equations, the following causal links may be singled out. Their description also allows a better understanding of the "vision" of the economy which underlies the model itself. Let us start with the real side.

The growth process is both export-led and expectations-led. Given foreign demand and prices, real exports grow according to domestic competitiveness and to supply constraints. Export growth enhances output growth which in turn modifies expectations and, consequently, real capital formation. Output growth also influences real imports, aggregate public consumption, direct taxes and the level of private consumption through the determination of disposable income. Changes in inventories, whose desired level is linked to expected output, act as a buffer in output determination.

The performance of real aggregates is also deeply influenced by price behaviour based on cost push and monetary mechanisms. Prices also enter in the determination of financial variables whose behaviour is closely connected with that of real variables.

A central place in the model is occuped by credit, whose expansion, as determined by the behaviour of banks, influences real capital accumulation as well as exports of goods and services and capital movements. An important role is also played by the rate of interest, for it influences the demand for money (and hence real consumption), credit expansion, and the accumulation of net foreign assets. The rate of interest is determined both by the foreign rate of interest and — given the demand for money — by the supply of money whose expansion is governed by the monetary authorities on the basis of their own policy targets and of the other channels of money creation (represented by the government and the balance of payments).

In sum, our model stresses real and financial accumulation in an advanced open economy in which aggregate demand and supply (and expectations) on the one hand and liquidity (i.e. money and credit availability) on the other play crucial roles.

2.2 The Single Equations

For convenience of exposition, the model is first laid out in tabular form and then each equation is commented on; for economy of notation, the disturbance terms are omitted and the model is described in deterministic terms. A hat ($\hat{}$) refers to the partial equilibrium level or desired value of the hatted variable, a tilde (~) to expectations; the symbol D denotes the differential operator d/dt, and log the natural logarithm. All variables are defined at time t. All parameters are assumed positive, unless otherwise specified.

Table 1 - Equations of the model

Private consumption

$$D\log C = \alpha_1 \log(\hat{C}/C) + \alpha_2 \log(M/M_d), \qquad (1)$$

where

$$\hat{C} = \gamma_1 e^{\beta_1 D\log Y} \left(\frac{P}{PMGS}\right)^{\beta_2} (Y-T/P), \beta_1 \gtreqless 0, \beta_2 \gtreqless 0; M_d = \gamma_2 i^{-\beta_3} TIT P^{\beta_4} Y^{\beta_5}, \beta_3 \gtreqless 0, \qquad (1.1)$$

Rate of growth in fixed capital stock

$$Dk = \alpha_3 [\alpha' \log(\hat{K}/K) - k] + \alpha_4 Da, \qquad (2)$$

where

$$\hat{K} = \gamma_3 \tilde{Y}, \qquad \gamma_3 = \kappa u, \qquad (2.1)$$

Expected output

$$D\log \tilde{Y} = n \log(Y/\tilde{Y}), \qquad (3)$$

Imports

$$D\log MGS = \alpha_5 \log(\hat{MGS}/MGS) + \alpha_6 \log(\hat{V}/V), \qquad (4)$$

where

$$\hat{MGS} = \gamma_4 P^{\beta_6} PMGS^{-\beta_7} Y^{\beta_8}, \qquad \hat{V} = \gamma_5 \tilde{Y}, \qquad (4.1)$$

Exports

$$D\log XGS = \alpha_7 \log(\hat{XGS}/XGS) - \alpha_8 Da, \qquad (5)$$

where

$$\hat{XGS} = \gamma_6 (PXGS/PF)^{-\beta_9} YF^{\beta_{10}} (\gamma_3 Y/K)^{-\beta_{11}}, \qquad (5.1)$$

Output

$$D\log Y = \alpha_9 \log(\tilde{Y}/Y) + \alpha_{10} \log(\hat{V}/V), \qquad (6)$$

Price of output

$$D\log P = \alpha_{11} \log(\hat{P}/P) + \alpha_{12} D'm, \qquad (7)$$

where

$$\hat{P} = \gamma_7 PMGS^{\beta_{12}} W^{\beta_{13}} PROD^{-\beta_{14}}, \qquad (7.1)$$

Price of exports

$$D\log PXGS = \alpha_{13} \log(\widehat{PXGS}/PXGS), \qquad (8)$$

where

$$\widehat{PXGS} = \gamma_8 P^{\beta_{15}} PF^{\beta_{16}}, \qquad (8.1)$$

Money wage rate

$$D\log W = \alpha_{14} \log(\hat{W}/W), \qquad (9)$$

where

$$\hat{W} = \gamma_9 P^{\beta_{17}} e^{\lambda_5 t}, \qquad (9.1)$$

Interest rate

$$D\log i_{TIT} = \alpha_{15} \log(M_d/M) + \alpha_{16} \log(\gamma_{10} i_f / i_{TIT}), \qquad (10)$$

Bank advances

$$D\log A = \alpha_{17} \log(\hat{A}/A), \qquad (11)$$

where

$$\hat{A} = \gamma_{11} i_{TIT}^{\beta_{18}} M, \qquad \beta_{18} \gtreqless 0, \qquad (11.1)$$

Net foreign assets

$$D\log NFA = \alpha_{18} \log(\widehat{NFA}/NFA) + \alpha_{19} Da, \qquad (12)$$

where

$$N\hat{F}A = \gamma_{12} i_{TIT}^{-\beta_{19}} i_f^{\beta_{20}} (PY)^{\beta_{21}} (PF \cdot YF)^{-\beta_{22}} Q^{\beta_{23}}, \qquad \beta_{23} \gtreqless 0, \qquad (12.1)$$

Monetary authorities' reaction function

$$Dm = \alpha_{20}(\hat{m} - m) + \delta_3 Dh + \delta_4 Dr, \qquad (13)$$

where

$$\hat{m} = \delta_1 \log(R/\gamma_{13} PMGS \cdot MGS) - \delta_2 (D\log P - D\log PF), \qquad (13.1)$$

Taxes

$$D\log T = \alpha_{21} \log(\hat{T}/T), \qquad (14)$$

where

$$\hat{T} = \gamma_{14}(PY)^{\beta_{24}}, \qquad (14.1)$$

Public expenditure

$$D\log G = \alpha_{22} \log(\gamma_{15} Y/G), \qquad (15)$$

Inventories

$$DV = Y + MGS - C - DK - XGS - G, \qquad (16)$$

International reserves

$$DR = PXGS \cdot XGS - PMGS \cdot MGS + (UT_a - UT_p) - DNFA, \qquad (17)$$

Fixed capital stock

$$D\log K = k, \qquad (18)$$

Rate of growth in bank advances

$$a = D\log A, \qquad (19)$$

Rate of growth in money supply

$$m = D\log M, \qquad (20)$$

Public sector's borrowing requirement

$$DH = PG - T, \qquad (21)$$

Rate of growth in H

$$h = D\log H, \qquad (22)$$

Rate of growth in international reserves

$$r = D\log R. \qquad (23)$$

Table 2 - Variables of the model[a]

Endogenous

A	= nominal stock of bank advances
a	= proportional rate of growth of A
C	= private consumption expenditure in real terms
G	= public expenditure in real terms
H	= public sector borrowing requirement
h	= proportional rate of change of H
i_{TIT}	= domestic nominal interest rate
K	= stock of fixed capital in real terms
k	= proportional rate of change of K
M	= nominal stock of money ($M2$)
m	= proportional rate of change of M
MGS	= imports of goods and services in real terms
NFA	= nominal stock of net foreign assets
P	= domestic price level
$PXGS$	= export price level
R	= nominal stock of international reserves
r	= proportional rate of change of R
T	= nominal taxes
V	= stock of inventories in real terms
W	= money wage rate
XGS	= exports of goods and services in real terms
Y	= real net domestic product and income
$\tilde{Y}$	= expected real net domestic product and income

Exogenous

i_f	=	foreign nominal interest rate
PF	=	foreign competitors' export price level
$PMGS$	=	import price level
$PROD$	=	labour productivity
Q	=	ratio of the forward to the spot exchange rate
t	=	time
$(UT_a - UT_p)$	=	net unilateral transfers, in nominal terms
YF	=	real world income

---

[a] Details of the variables and sources of data are given in Appendix II, §II.1.

2.2.1 The Real Sector

In equation (1), *real private consumption* C adjusts to its desired level, $\hat{C}$, which is given by the average propensity to consume applied to real disposable income ($Y-T/P$). The propensity to consume, in turn, is a function of other variables. Firstly, it may vary over the trade cycle, that is, according to the rate of growth of Y, either pro-cyclically (β_1 positive) or anti-cyclically (β_1 negative); this second possibility arises if a "ratchet effect" is operative.

Secondly, it may vary with the terms of trade. The inclusion of the terms of trade effect was advocated 35 years ago in the seminal contributions by Laursen and Metzler (1950) and Harberger (1950)[1]. Strictly speaking, the Laursen and Metzler reasoning (1950, p. 286), which leads to the assumption $\beta_2 < 0$, is applicable only to a model, like theirs, in which the domestic price level is assumed to be constant. If both P and $PMGS$ vary, the outcome cannot be assessed on *a priori* grounds; therefore the sign of the final effect of the terms of trade on consumption is ambiguous.

The second term in equation (1) represents the effect on consumption

---

[1] For a recent theoretical study of this effect see Svensson and Razin (1983). The empirical evidence is examined in Deardoff and Stern (1978a) and in Obstfeld (1982).

of an excess supply of money. This has a twofold interpretation: a) the speed at which C adjusts to $\hat{C}$ is not a constant but is an increasing function of the excess supply of money; a suitable linear approximation then gives eq. (1). b) the private sector reacts to an excess supply of money both by inducing changes in the interest rate — see eq. (10) — and by adjusting consumption; therefore, we have a spillover effect on consumption of monetary disequilibrium. The demand for money in nominal terms M^d is a function of the interest rate and of nominal income. Traditional demand-for-money theory requires that the sign of the interest rate elasticity (β_3) be negative. In our case, however, since the variable employed to represent the money stock ($M2$) includes bank deposits whose demand is positively related to the rate of interest on these deposits (a rate which in Italy is closely related to i_{TIT}) the sign of β_3 could be either positive or negative on a priori grounds. The price level P and real output Y, whose product constitutes nominal income, appear with separate elasticities to allow for both a non-unitary elasticity with respect to income (economies of scale etc.) and a possible presence of money illusion (which would imply that the demand for money in real terms is not independent of P).

The supply of money in nominal terms, M, is determined by the monetary authorities' reaction function — see eqs. (13) and (20). Note that it makes no difference whether monetary disequilibrium is expressed in nominal or in real terms, for $(M/P)/(M^d/P)=M/M^d$.

The *investment function* implicit in eq. (2) is a development of the capital stock adjustment principle. According to the traditional version of this principle, the rate of capital formation is a function of the gap between desired capital stock $\hat{K}$ and actual capital stock K, that is [2]

$$D\log K = \alpha' \log(\hat{K}/K), \qquad \alpha' > 0,$$

---

[2] The assumption that the independent variable in the investment function is not I but I/K has a pedigree that dates at least as far back as 1933: see — though in another context — Kalecki (1933); more recent references are Bergstrom and Wymer (1976), and Knight and Wymer (1978).

where $\hat{K}$ depends on output. In our opinion the gap between $\hat{K}$ and K does not give rise directly to actual capital formation, but is rather the determinant of the *desired* rate of capital formation $\hat{k}$, namely

$$\hat{k} \equiv \widehat{D\log K} = \alpha' \log(\hat{K}/K).$$

We then assume that the relevant economic agents adjust the actual rate of capital formation $k \equiv D\log K$ to the desired rate according to the partial adjustment equation

$$Dk = \alpha_3(\hat{k}-k) = \alpha_3[\alpha'\log(\hat{K}/K) - k],$$

which is equation (2). Therefore the capital stock changes according to a second-order adjustment process.

A second modification is that instead of relating $\hat{K}$ to current output, we think it more plausible to relate it to expected output $\tilde{Y}$, whose determination is discussed below. The relation between $\hat{K}$ and $\tilde{Y}$ is given by $\gamma_3 = \kappa/u$, that is the ratio between the capital/output ratio and the desired degree of capacity utilization.

The second term can be interpreted as the result of the linearization of the speed of adjustment $\alpha_3 = \phi(Da)$, $\phi' > 0$; that is to say, a faster rate of increase in the flow of bank advances, which is determined by bank behaviour — see eqs. (11) and (19) — has a favourable effect on the speed of adjustment under consideration, for it makes the implementation of investment decisions easier. In other words, the inclusion of credit availability in the investment function takes account of what Keynes (1937) termed the "finance motive" in the demand for liquidity by entrepreneurs, according to which liquidity shortage (i. e. credit rationing) could — *ceteris paribus* — hinder the implementation of *decided* investment.

It should be pointed out that, since reliable data are not available on the public sector's net investment expenditure, DK is net capital formation of the whole economy.

Equation (3) expresses the *formation of expectations* on output according to an adaptive mechanism[3], where $\eta > 0$ is the coefficient of expec-

---

[3] See above, section 1, and below in the text, where eq. (6) is examined. We do

tations.

The specification of the behaviour of *real imports* for goods and services (MGS) in eq. (4) has two determinants. First of all, imports adjust to their desired value $\widehat{MGS}$, which is a function of home and foreign prices as well as of domestic output. This is the usual specification of the (desired) demand for imports, with one important modification: instead of considering relative prices ($PMGS/P$), we consider the two absolute prices separately. In fact, previous research, though in a single-equation context, points to the possibility of money illusion in the demand for imports (Gandolfo, 1979); therefore, the use of relative prices — which is warranted only if absence of money illusion is assumed — would be a misspecification.

The second determinant is due to the adjustment of inventories V to their desired level $\widehat{V}$. The change in the level of inventories is a residual in the goods market (see eq. (16) in table 1, and below in the text), but it is assumed that the relevant economic agents have a desired ratio γ_5 of inventories to expected output[4] and that they will increase both output — see below, eq. (6) — and imports whenever inventories are below the desired level. Thus, inventories act as a buffer between supply and demand in the goods market. Furthermore, the presence of inventories in the import adjustment equation might also reflect "speculative" considerations, and this also would justify a different speed, whence the separate consideration of inventories instead of their inclusion in $\widehat{MGS}$ as a further explanatory variable.

---

not wish to enter here into the controversy concerning rational vs. adaptive (or other types of) expectations. Arguments in favour of adaptive expectations (including the cases in which, under general realistic circumstances, the "rational" expectation-formation process is an adaptive mechanism) can be found, e.g., in B. Friedman (1979), Gordon (1976), Granger and Newbold (1977), Knight (1976), Malinvaud (1982), Mussa (1978), Muth (1961), Sargent and Wallace (1973), etc.

[4]That is, the *desired* level of inventories is related to *expected*, and not to actual, output. The reason for this is the same as that which prompted us to assume that the desired stock of capital is related to expected output rather than to actual output. As we take it that entrepreneurs look ahead into the future, the level to which these stocks are being adjusted by them are, more plausibly, related to expected output.

Real exports of goods and services — eq. (5) — adjust to their partial equilibrium level, $\widehat{XGS}$, which is determined by foreign demand for exports and by a supply constraint. More precisely, foreign demand depends in the usual way on relative competitiveness[5] (given by the ratio of domestic export prices, *PXGS*, to foreign competitors' export price in world markets, *PF*), and on world income *YF* (which can be considered by itself or as a proxy for world demand). Given the effect of foreign demand, the partial equilibrium level of exports also reflects the influence of domestic supply represented by the deviations of the output/capital ratio from its desired value $(1/\gamma_3)$; these can be considered as a proxy for the degree of capacity utilization[6]. In other words, as the utilization of productive capacity increases, the negative impact of a supply constraint will be felt on exports.

The inclusion of the *variation in the rate of change of bank advances* is due the observation that, whenever a credit squeeze occurs, Italian producers try to increase the expansion of exports by every means[7]. The reason (De Cecco, 1967, p. 54) is that by so doing they increase their availability of liquidity notwithstanding the credit squeeze: in fact, either exports are paid for in cash by the foreign buyers, or the exporters' credit is immediately financed by banks (which, on the contrary, do not easily finance the credit for sales on the domestic market because of the credit squeeze).

Eq. (6) specifies the behaviour of the *supply side*, that is of real output, according to two components. Firstly, producers adjusts current output towards expected output [while so doing, they might run into capacity problems, which are dealt with in eq. (2)]. Secondly, as al-

---

[5] Since $\widehat{XGS}$ is not determined by a pure demand equation but by a mongrel equation containing supply elements as well, the presence or absence of money illusion cannot be detected, so that we have not modified the traditional relative prices variable.

[6] In fact, $Y/[(1/\gamma_3)K]$ can be considered as the ratio of actual to desired capacity output.

[7] Including all kinds of "non-price" competition. These, therefore, although not explicitly included in the export equation, are indirectly accounted for.

ready observed in relation to eq. (4), although the level of inventories is a residual in the goods market, we assume that producers have a desired ratio (γ_5) of inventories to expected output and that they will increase output (as well as imports) whenever inventories are below their desired level. This is the role of inventories as a buffer between supply and demand in the goods market. Furthermore, inventories have another fundamental role to play in eq. (6). In fact, if they were absent there, then — given the adaptive expectations mechanism defined in eq. (3) — equations (3) and (6) would form a self-contained sub-model which would determine Y and $\tilde{Y}$ independently of the rest of the model. This undesirable dichotomy is avoided by the presence of inventories in eq. (6), which link Y and $\tilde{Y}$ to the rest of the model through eq. (16).

We wish to stress here the fundamental role of expected output in the model. In addition to being a determinant of the supply of current output, it is also a determinant of the accumulation of real stock (of both fixed capital and inventories), thus capturing the crucial role played by the behaviour of firms in an open economy.

We now turn to the price sector of the model.

2.2.2 The Price and Wage Sector

The *domestic price level P* — eq. (7) — adjusts according to both cost push and monetary factors. The latter, however, do not enter into the determination of the partial equilibrium level $\hat{P}$, but rather influence the speed at which the actual price level adjusts to $\hat{P}$. We assume that this speed is an increasing function of the "easiness" of monetary conditions (as represented by Dm); a suitable linear approximation then gives eq. (7).

The determinants of $\hat{P}$ are both domestic and foreign. The latter are represented by import prices[8], capturing both the effect of a rise in

---

[8] For a detailed analysis of the effects of foreign prices on domestic price determination see Deardoff e Stern (1978b).

the cost of imported factors of production and a possible "foreign competitiveness effect", which induces domestic producers to take into account the prices of competing imported goods when they determine the prices of their products. The push from domestic costs is represented both by the level of the nominal wage rate W and, with an inverse relationship, by the level of productivity $PROD$ (exogenously given)[9]. We allowed for separate influences of wages and productivity rather than considering the familiar ratio $W/PROD$, because they might each influence $\hat{P}$ differently: if there are frictions, non-competitive situations etc., the same proportional increase in W and in $PROD$ does not necessarily leave $\hat{P}$ unaltered.

The *nominal wage rate*, in turn — eq. (9) — adjusts to a partial equilibrium level $\hat{W}$, which is a direct function of the domestic price level. Institutional factors (as for example a dominant "social pressure" — in Hicks' sense, 1974, ch. III — for rising wages, etc.) suggest for many industrial countries that target nominal wages exceed a level determined only by the domestic price behaviour. These factors have been taken into account by the introduction of a trend among the determinants of $\hat{W}$[10]. Eqs. (7) and (9) considered together represent a wage-price spiral situation and take account, albeit approximately, of the wage indexation effect for the Italian economy.

The introduction of a separate equation for *export prices* — eq. (8) — is justified by the consideration that exporting firms fix prices by taking account not only of the elements which enter into the determination of domestic prices (though with a possibly different weight, whence the presence of the parameter β_{15}) but also of the foreign competition barrier. This is represented by PF, as defined in the export equation, which has a positive effect on the level of $\hat{PXGS}$.

---

[9] We tried to make $PROD$ endogenous (for example by using a technical progress function à la Kaldor), but estimation results were never satisfactory. See Appendix II, §II.2.

[10] It is often assumed that this trend coincides with the trend of productivity. But, in our opinion, such an assumption is not necessarily warranted by the institutional factors etc., which underline the trend in question. Therefore we allowed the trend in W to be different from that in $PROD$.

Since export prices are charged by domestic producers in foreign markets, there is no reason to believe — contrary to the domestic price level adjustment equation — that the speed of adjustment is a function of a domestic monetary variable. It could, at most, be a function of a world monetary variable, which, however, would more realistically operate on *PF*.

2.2.3 The Financial Sector

Equation (10) determines the rate of change of i_{TIT} which, as we said above, is the representative interest rate in our model. Therefore, the structure of interest rates is not considered. This simplification can be justified on the basis of the fact that a complete modelling of this structure would have required a large increase in the size of the model, which we wished to avoid. It should also be noted that, owing to the lack of reliable data on short-term interest rates for a sufficiently long period, in Italian empirical analyses i_{TIT} is often used as a proxy for the interest rate on bank advances.

The rate of change of the interest rate is determined by two different effects. The first is a domestic effect and is represented by the traditional excess demand for money whose specification has already been discussed in relation to the consumption function. The second effect accounts for the influence of the foreign interest rate (also in this case i_f — the US Treasury bill rate — is considered as a "representative" rate). The idea is that, given the international "financial dependence" of the Italian economy, whenever the foreign interest rate rises (less so when it falls), the domestic rate sooner or later has to be varied accordingly, as a policy measure, in order to avoid unwanted capital outflows (as is specified in eq. (12)[11]). Therefore eq. (10) has a hybrid nature, for both "market" and "policy" forces come into play, as is the case for the Italian economy.

---

[11] In fact, one of the reasons which compelled the Italian monetary authorities to abandon the pegging of the domestic interest rate (in 1969) was the outflow of capital determined by the increasing differential in favour of foreign interest rates.

Equation (11) describes the behaviour of the credit system in determining bank advances. A full treatment of the credit market would require the specification of both demand and supply of credit, but dimension limitations have forced us to include only one equation, despite the central role played by credit availability in our model. According to our specification, bank advances A are "supply determined", i.e. determined by the behaviour of the banking sector. Although this assumption might be considered over-simplistic, it is consistent with the idea that in Italy credit rationing conditions usually prevail in the market for bank advances so that supply generally falls short of demand. The desired level of bank advances $\hat{A}$, as a proportion of the scale variable M (which in turn can be considered as a proxy for bank deposits, and these are, as we said, the greatest component of M) depends on i_{TIT}.

As regards i_{TIT}, its "catch-all" nature causes the *a priori* indeterminateness of its effect on $\hat{A}$. In fact, this effect should be positive if i_{TIT} is a proxy of the interest rate on advances, and it should be negative if i_{TIT} is a proxy of the interest rate on alternative uses of funds by banks (one such use is investment in government bonds), for in the latter case i_{TIT} is an opportunity cost.

The *net outflow of capital* — eq. (12) — is explained by the adjustment of the actual to the desired stock of net foreign assets. Hazardous as it may be, this equation is meant to explain all kinds of capital movements (both long-term and short-term), so that the explanatory variables present in $\widehat{NFA}$ are of both a long-term and a short-term nature. To be precise, the desired stock of net foreign assets depends on two scale variables, namely on domestic money income (in a positive way) and on foreign money income (in a negative way)[12]. Furthermore, $\widehat{NFA}$ is a decreasing function of the home-abroad interest-rate differential.

---

[12] If there were no money illusion, both β_{21}, and β_{22} ought to be equal to 1. But we do not believe that there are good enough reasons for assuming absence of money illusion *a priori*, so that we did not introduce the constraints $\beta_{21} = \beta_{22} = 1$.

The presence of the variable Q, defined as the ratio of the forward to the spot exchange rate, may be justified according to two different explanations due to the hybrid nature of eq. (12). If we assume that our representative rates i_{TIT} and i_f are well suited to take account of interest rate differentials, then Q should be considered as a proxy for exchange rate expectations[13]. If this is not the case then Q itself could be considered as a proxy for interest rate differentials. As a consequence, the sign of β_{23} may be either positive or negative[14]. The indeterminateness of the sign of β_{23} is again due to the twofold nature of i_{TIT}. If i_{TIT} is to be considered a proxy for a short-term interest rate, then Q should account for exchange rate expectations, and so $\beta_{23} > 0$, for the expectation of a depreciation implies a greater desired stock of assets abroad. On the contrary, if i_{TIT} is to represent the long-term interest rate (and i_f the foreign long-term interest rate), then Q should represent the short-term interest rate differential.

The inclusion of the term $\alpha_{19} Da$ may be justified on similar grounds as those discussed in relation to eq. (2). A faster rate of increase in the flow of bank advances enhances (the speed of adjustment of) the accumulation of net foreign assets.

---

[13] Since the exchange rate is not included in the model as an endogenous variable (see below), no endogenous exchange rate expectations were inlcuded.

[14] The well-known $IRPT$ condition yields $FR/E = (1+i_h)/(1+i_f)$, where FR = forward exchange-rate, E = spot exchange-rate (both defined as units of domestic currency per unit of foreign currency), i_h = domestic interest rate, i_f = foreign interest rate. Consequently, $Q \gtreqless 1$ according as $i_h \gtreqless i_f$, and the ratio Q can be taken as a proxy for the interest differential. We assume that the more favourable (unfavourable) to the home country the interest differential is — namely the greater (smaller) Q is — the smaller (greater) the desired stock of net foreign assets will be. We are well aware that divergencies from $IRPT$ may (and do) occur (see, for example, Frenkel and Levich, 1979; Solnik, 1979), but we do not think that these invalidate the use of Q as a rough proxy.

2.2.4 Policy Reaction Functions

Equation (13) is a reaction function of the monetary authorities. We have followed this "policy reaction function" approach (in the line of the feedback policy approach pioneered by Phillips, 1954), instead of the optimizing approach, because our model purports to reflect the *actual* behaviour of the Italian monetary authorities, who do not optimize an objective function but behave following feedback rules. We are well aware that:

i) on the one hand, feedback policy rules can be derived from a quadratic-linear optimisation problem (quadratic objective function subject to a linear econometric model: see, for example, Theil, 1964, ch. 2; or — in the dynamic optimisation context — quadratic objective function subject to a linear differential equation system giving the equations of motion of the variables: see, for example, Athans and Falb, 1966, ch. 9, Aoki, 1976, ch. 5);

ii) on the other hand, the weights of an "unconscious" objective function can, in principle, be derived from the observed behaviour of the policy authorities (see, for example, Pissarides, 1972; Makin, 1976; Chow, 1981) or by using what we would call a "maieutic (Socratic) method" (such as that suggested, for example, by Rustem, Velupillai, and Westcott, 1978; or that suggested by the interactive multiobjective optimization approach described for example in Gruber, ed., 1983);

but we had to leave this for future research. Therefore we shall examine the effects of alternative policies by using the traditional simulation approach which, however, is particularly well suited for the Italian case, where a great number of disparate policies have been suggested (for example for coping with the problem of inflation) by several different sources (private researchers, trade unions, political parties, research institutions, etc. etc.) with no optimising basis and with usually unproved assertions on their efficacy. The simulation of these policies by using the same model will make it possible, at least, to get a clear and unprejudiced idea of their presumable overall effects.

Going back to eq. (13), it should be observed that the use of the proportional rate of growth of M_2 as "the" policy variable is imprecise for obvious reasons. However, since a fully fledged analysis of the monetary structure of the Italian economy, which would have allowed us to determine all policy variables (as well as intermediate and final targets) exactly, would have required too large an extension, the choice was almost obligatory.

The choice fell on a "quantity" variable (although "price" policy is in part included in eq. (10)) for both theoretical and empirical considerations. In fact, "quantity" effects (such as rationing, in the wide sense, both by monetary authorities and by the banking sector) have often been more important than "price effects". The change in the proportional rate of change of the money stock rather than the money stock itself seemed the most suitable variable, because a "restrictive" monetary policy is usually considered to be one which causes a slowing down in the proportional rate of increase of the money stock (and not, or not necessarily, an absolute decrease in the level of this stock); a similar consideration holds for an "expansionary" monetary policy.

Thus our choice is different from that suggested by Black (1983), according to whom (p. 190) "Despite the frequent use by economists of market determined variables, such as the rate of growth of the money supply or short-term interest rates, as "instruments" of central banks, I believe that the term should be limited to variables which are actually in the hands of the authorities". These considerations lead him to use the discount rate as the dependent variable in the monetary authorities' reaction function. Now, apart from institutional considerations (the discount rate is not a policy instrument on which the Bank of Italy relies very much), we believe that the use of the rate of growth of the money supply can be justified on theoretical grounds if properly treated, namely if the distinction is made between money creation by the central bank (a variable actually in the hands of this policy authority) and money creation through other channels. Now the idea behind eq. (13) is the following. Let DM_{cb}, DM_g, DM_{bp} denote money creation (or destruction) respectively by the central bank, the government, the balance of payments[15]. Therefore

$$DM = DM_{cb} + DM_g + DM_{bp},$$

whence

$$\frac{DM}{M} = \frac{M_{cb}}{M}\frac{DM_{cb}}{M_{cb}} + \frac{M_g}{M}\frac{DM_g}{M_g} + \frac{M_{bp}}{M}\frac{DM_{bp}}{M_{bp}},$$

and, letting lower-case letters denote proportional rates of change,

$$m = \mu_{cb} m_{cb} + \mu_g m_g + \mu_{bp} m_{bp},$$

where of course the weights (the μ's) add up to one. Now, the central bank has an overall desired value of m, say $\hat{m}$ (which depends on the central bank's targets in a way to be specified later), and we assume that it can control m_{cb} but cannot directly and immediately control m_g and m_{bp}[16]. Therefore the central bank determines a target or desired value of m_{cb} (denoted by $\hat{m}_{cb}$) as that value which, given the actual values of m_g and m_{bp} and the weights, corresponds to the target $\hat{m}$, namely

$$\mu_{cb}\hat{m}_{cb} = \hat{m} - \mu_g m_g - \mu_{bp} m_{bp}. \qquad (*)$$

Now, due to inevitable frictions, lags etc., the actual value of m_{cb} cannot be maintained of brought to its desired value $\hat{m}_{cb}$ instantaneously, but is manoeuvered according to the partial adjustment equation

$$Dm_{cb} = \alpha_{20}(\hat{m}_{cb} - m_{cb}).$$

---

[15] Each of these "agents" has its channels, *modus operandi*, etc., but we do not model them for we are interested in the final result only.

[16] This is a simplifying assumption, for the central bank may have a certain amount of control on m_g and m_{bp} according to the different institutional settings. To avoid confusion, we must however recall that by "sterilization" of balance of payments flows we mean the fact that the effect on domestic liquidity of these flows is offset by opposite changes in the variables directly controlled by the central bank (m_{cb} in our scheme). This is not a control on m_{bp} but is just an offsetting action; a control on m_{bp} would imply specific administrative measures. See Herring and Marston (1977).

If we multiply through by μ_{cb} we have

$$\mu_{cb} Dm_{cb} = \alpha_{20}(\mu_{cb}\hat{m}_{cb} - \mu_{cb}m_{cb}),$$

and by using eq. (*) we obtain

$$\mu_{cb} Dm_{cb} = \alpha_{20}[(\hat{m} - \mu_g m_g - \mu_{bp}m_{bp}) - \mu_{cb}m_{cb}]$$

$$= \alpha_{20}(\hat{m} - m).$$

The addition of $(\mu_g Dm_g + \mu_{bp} Dm_{bp})$ to both sides, and the further assumption that the weights are approximately constant, yield

$$Dm = \alpha_{20}(\hat{m} - m) + \mu_g Dm_g + \mu_{bp} Dm_{bp}.$$

Now, m_g depends on the rate of change in the government's deficit (or, more generally, the public sector borrowing requirement[17]) and m_{bp} depends on the rate of change in international reserves, so that, given the definitions (22) and (23),

$$Dm = \alpha_{20}(\hat{m} - m) + \delta_3 Dh + \delta_4 Dr,$$

where the parameter δ_3 incorporates μ_g as well as the parameter that links m_g to h, and δ_4 incorporates μ_{bp} and the parameter that links m_{bp} to r[18]. This gives eq. (13) of the model.

It remains to determine $\hat{m}$, and this presents quite a problem. In fact, for several reasons — the lack of an effective fiscal policy being the main one — the monetary authorities have been burdened by a large number of targets for them to take care of, although obviously not always at the same time and with the same weights.

Now, as we said above, our equation purports to reflect the actual behaviour of the Italian monetary authorities over the sample period, namely to describe what they did, not what they ought to have done (following the modeller's or someone else's opinion[19]). Although, as we said above, the Italian central bank has occasionally taken several

---

[17] It goes without saying that a precise modelling of this relationship would require the modelling of the ways of financial the government deficit (sales of bonds and bills to the public, to the banking sector, to the central bank etc. etc.), but for the reasons already stated we did not wish to add these complications.

[18] This parameter includes administrative measures, if any.

[19] For a description of Italian monetary policy the best source is the Banca d'I-

targets into account (among which capital accumulation, real output growth, employment), we must try to include in the policy function only those targets which seem to have been a constant concern [20].

An element to which the Bank of Italy has always paid a great deal of attention is, avowedly, the "months of financial covering of imports", that is how long the current flow of imports can be maintained if the existing stock of reserves are used up for this purpose; this implies the existence of a desired ratio (γ_{13}) of R to the value of imports. The consequent behaviour is to follow a restrictive policy whenever the actual ratio is lower than the desired ratio, and viceversa. This gives the first term in eq. (13.1).

A second — and probably even more important — element is the anti-inflationary target of monetary policy, which however does not aim at absolute, but at relative price stability. Put differently, the monetary authorities' target rate of inflation is not zero, but is the foreign rate of inflation, proxied by the proportional rate of increase in PF. The fact that the monetary authorities, *ceteris paribus*, decrease m to cope with, say, too high a rate of inflation, reflects their opinion that this decrease should have an effect on the rate of growth of P. This opinion, generally shared by the monetary authorities is (in part) validated by eq. (7).

Let us now consider the fiscal sector. The fact that no consistent and effective fiscal policy seems to exist in Italy led us to drop any idea of introducing fiscal policy reaction function(s), and to model tax revenues and government expenditure in a very simple way.

---

talia Annual Report (an abridged English version is available from the Banca d'Italia), where the targets are usually clearly stated in the Governor's Concluding Remarks. See also, for the period under consideration, Allen and Stevenson (1974, ch. 5), Baffi (1971), Fazio (1979), Hodgman (1974, ch. V), OECD (1973), Caranza and Fazio (1983).

[20] The inclusion of all the targets would make it impossible to disentangle their weights unless these were made variable in time (a method could be that of the "switching function(s)" suggested by Wymer, 1979, 1983), but such a complication is outside the scope of the present work. As a matter of fact, when we included further targets in previous versions of the model, the results were always poor.

In eq. (14), total nominal taxes (net of transfers) adjust to a level which depends on nominal income. Note that β_{24} should be greater than one, to reflect the fiscal drag, if T included direct taxes only. But, since T includes all kinds of taxes, many of which are proportional, or less than proportional, to nominal income, we way assume $\beta_{24} \leqslant 1$.

In eq. (15), real government expenditure adjusts to a value which is a fraction of real output; γ_{15} may be interpreted as a "desired" ratio of G to Y. This is consistent with the idea that G is not used for policy purposes, but does not prevent us from introducing an *a priori* given policy function in simulation exercises. Note that G does not include government investment expenditure, which is included in eq. (2).

Eqs. (16)-(23) are definitional equations.

2.2.5 Final Considerations

We should now notice that the exachange-rate is not explicitly present as an endogenous variable in the model. It is, of course, indirectly present through Q and the two foreign price variables (*PMGS* and *PF*): the last two are defined in domestic currency terms, but they could be defined as follows:

$$PMGS = E \cdot PMGS_f, \qquad PF = E \cdot PF_f,$$

where E is the exchange rate (defined as units of domestic currency per unit of foreign currency) and the subscript f denotes magnitudes in foreign currency.

As our sample period covers both fixed exchange rates and managed float, the fact the exchange rate is not considered explicitly reflects the assumption that the behavioural functions present in the model have not been significantly altered by the change-over from the fixed exchange rate regime to the managed float [21]. If the empirical analysis

---

[21] In other words, and by way of example, economic agents react to *PMGS* as such, rather than to E and $PMGS_f$ separately.

rejects this assumption, then the switch in the exchange rate regime will have to be dealt with. This can be accomplished either by introducing dummy variables or by means of more sophisticated methods, such as the switching function suggested by Wymer (1979, 1983) as a general method of coping with models which contemplate two different states of the economy, each described by different behavioural functions.

We can anticipate that the empirical analysis (chapter 3) did not reject our assumption. Therefore the model can also be used for simulations which involve the exchange rate; in particular, an equation for the endogenous determination of the exchange rate can be added in simulation. As is well-known, there are many competing theories for the determination of the exchange-rate: *PPP*, traditional "flow" hypothesis, monetary approach (various versious), asset market approach (various versions), etc. (for surveys see Isard, 1978; Krueger, 1983; Horne, 1983). The presence, in our model, of both flow and stock variables (real and financial), which interact in a disequilibrium setting, and the possibility of determining econometrically the adjustment speeds of all the variables involved, will allow us to examine the determination of the exchange rate with an unprejudiced eye (see chapter 4, section 4.3).

It is now useful to discuss briefly the main differences between the most important continuous models and our own. Both the Knight and Wymer (1978) and the Australian models (RBA, 1977; Jonson and Trevor, 1981) include a *labour market* which is modelled in terms of aggregate demand (derived from a neoclassical aggregate production function) and supply of labour. No such analysis is included in our model for several reasons, among which the fact that the Italian labour market does not conform to this theoretical representation; an adequate treatment would have required too large an increase in the number of equations.

In most other models the real and financial sectors are linked, in a somewhat traditional way, only through the demand for money and the interest rate entering the consumption (and, in some cases, the investment) function. In our case real and financial variables are much more closely interlocked, given also the central role of credit in the accumulation process. Our work also differs in this respect from the continuous models of the Italian economy produced by Chiesa (1979) and Tullio (1981), who follow the lines of the authors cited above. As a matter of fact both Chiesa and Tullio present a strict neoclassical-monetarist description of the Italian economy which is contrary to our eclectic view. A detailed comment on Tullio's model is contained in Gandolfo and Padoan (1982c).

2.3 Steady-State, Comparative Dynamics, Stability, and Sensitivity.

Since this is a medium-term model, it is important to check that it possesses an economically meaningful *steady-state solution*. This also helps as a check on the mathematical consistency of the model itself and as a source of cross-equation restrictions on the parameters.

The details of the procedure are treated in appendix I; here we state and comment on the results.

The steady-state growth rates of the endogenous variables are summarized in table 3, where λ_1 = growth rate of *PMGS* , λ_2 = growth rate of *PF*, λ_3 = growth rate of *YF*, λ_5 = time trend in *W*.

Table 3 - Steady-state growth rates

Variable	Growth rate
$C, K, \tilde{Y}, MGS, XGS, Y, V, G, T/P$	$\beta_9(\lambda_2 - \lambda_1) + \beta_{10}\lambda_3$
$P, PXGS$	λ_1
W	$\beta_{17}\lambda_1 + \lambda_5$
M, A	$\beta_4\lambda_1 + \beta_5[\beta_9(\lambda_2 - \lambda_1) + \beta_{10}\lambda_3]$
R, NFA, H, T	$\lambda_1 + \beta_9(\lambda_2 - \lambda_1) + \beta_{10}\lambda_3$
k, a, m, h, r, i_{TIT}	0

The real variables (including expected output[22]) grow at a rate which, given the elasticities, depends on the rest of the world variables, namely on the rate of growth of world output (term $\beta_{10}\lambda_3$), account being taken of the behaviour of the home country's competitiveness [term $\beta_9(\lambda_2-\lambda_1)$][23]. This is consistent with the "export led" way of looking at growth.

Domestic prices and export prices grow at the same rate, equal to the growth rate of foreign prices (represented by import prices). Although the equations for the endogenous determination of domestic and export prices are different, so that in disequilibrium (which is the "normal" state) their paths are not the same, in steady-state all the items in the balance of payments must grow at the same rate. Therefore the value of exports and the value of imports must grow at the same rate, and this means that export prices have to grow at the same rate as import prices, for the quantities of exports and imports both grow at the common rate of the real variables.

The nominal wage rate grows at a rate which is given by the sum of the indexation effect ($\beta_{17}\lambda_1$) and of the exogenous trend.

The stock of reserves R must grow at the same rate as the value of imports so as to maintain the desired ratio γ_{13}. Therefore, balance of payments equilibrium does *not* occur but, on the contrary, a constantly growing overall surplus takes place. This is an obvious consequence of the fact that — according to eq. (17) — an increase in reserves can be obtained only through a balance of payments surplus[24].

---

[22] In steady-state, expected output must grow at the same rate as actual output, but the former is not equal to the latter. In fact, as we show in appendix I, $\tilde{Y}^* \neq Y^*$ and, precisely $\log \tilde{Y}^* = \log Y^* - \rho_3/\eta$. Therefore, in steady-state $\tilde{Y}$ is always smaller than Y, and this is the mechanism which engenders the growth of output. In fact, as $\tilde{Y} < Y$, expectations are continually revised upwards, according to eq. (3) and, consequently, production is increased, according to eq. (6). This allows output to increase to match the increase in demand.

[23] Note that if we impose the further condition that home prices and foreign competitors' prices on the world market grow at the same rate, then obviously this term disappears. But this condition is not required for the existence of the steady-state.

[24] This may raise problems of consistency at the world level (which can be ignored

The stock of net foreign assets also grows at the common rate of growth of the items in the balance of payments, for this is the rate at which its derivative (that is, net capital outflows) grows.

It is interesting to note that the other financial variables (M and A) grow at a different rate from R and NFA. In fact, the rate of growth of the money supply (to which the growth rate of bank advances must be equal) is a weighted average of the rate of growth of prices and of the rate of growth of the real variables. This is a consequence of the presence of the elasticities β_4 and β_5 in the demand for money (which in steady-state equals the money supply). If we wish all the financial variables to grow at the same rate, we must impose the additional constraint $\lambda_1 + [\beta_9(\lambda_2 - \lambda_1) + \beta_{10}\lambda_3] = \beta_4\lambda_1 + \beta_4[\beta_9(\lambda_2 - \lambda_1) + \beta_{10}\lambda_3]$.[25] But this is not required for the existence of the steady-state.

It is important to notice that our model possesses a steady-state even *without* the imposition of some dubious constraints often found in this type of model, an example of which is that, in steady-state, real wages grow at the same rate as productivity, so that the distribution of income is constant. This would imply $\beta_{17}\lambda_1 + \lambda_5 = \lambda_1 + \lambda_4$, whence $\beta_{17} = 1 + (\lambda_4 - \lambda_5)/\lambda_1$ is completely determined. In our model, distribution will shift in favour of, or against, wage earners according to $(\beta_{17} - 1)\lambda_1 - \lambda_4 + \lambda_5 \gtreqless 0$.

As we said above, the solution for the steady-state usually implies restrictions on the parameters (some of which are of the cross-equation type). These are given in appendix I.

The *comparative dynamics*[26] of the steady-state growth is immediately

---

in our "small-country" model), for not all countries can have a permanent surplus. A generalized desire for a growing stock of reserves can be met, for example, if there is a reserve currency, and international liquidity is created by the deficit of the reserve currency country.

[25]A particular case of this constraint would be $\beta_4 = \beta_5 = 1$. But, for the reasons expounded in §2.2.1 when commenting on eq. (1), we do not believe this constraint to be justified on theoretical grounds.

[26] Since we deal with comparative dynamics, in what follows expressions such as "a higher value of x implies a lower value of y" or "an increase in x causes a decrease in y" etc., should be interpreted as meaning "consider a steady-state in which the value of x is higher than in another steady-state; then the value of y in the former steady-state is lower than in the latter".

obtained from the results given in table 3. For example, a higher growth rate of import prices implies a lower growth rate of the real variables and a higher growth rate of the price and wage variables. On the contrary, higher growth rates of foreign competitors' prices and of world income imply a higher growth rate of the real variables and have no effect on the price and wage variables. These results are quite obvious and are consistent with the "export-led" nature of our model. In fact, the growth rate of domestic prices and of export prices is the same as the growth rate of import prices. An increase in the growth rate of export prices reduces the competitiveness of home products on foreign markets, hence the lower growth rate of exports and of all the other real variables. An opposite effect occurs when the growth rate of foreign competitors' prices increases; also an increase in the growth rate of world income has a favourable effect on exports.

Similarly, a higher elasticity of exports with respect to relative prices (β_9) and with respect to world income (β_{10}) has a favourable effect on the growth rate of the real variables.

Finally, as regards *stability analysis* and *sensitivity*, these require the previous linear approximation of the model about the steady-state. These are purely technical matters and are worked out in appendix I. Once this linear approximation has been performed, however, no purely qualitative results can be obtained from a 23x23 differential equation system. Therefore the stability and sensitivity analyses have been postponed till the following chapter, after the numerical estimates of the parameters have been obtained; here we will only make a few general considerations. The importance of stability analysis has already been stressed in chapter 1. As regards sensitivity, it should be clarified that by *sensitivity analysis* we mean the analysis of the effects of changes in the parameters on the characteristic roots of the model. This can be performed in a general way by computing the partial derivatives of these roots with respect to the parameters, $\partial \mu_j / \partial \theta_i$, where μ_j is the j-th characteristic root and θ_i is the i-th parameter. This is very important both at the theoretical level and for policy purposes.

At the theoretical level, the analysis of the qualitative properties of the model (such as structural stability, bifurcations etc.) is greatly facilitated. A system of differential equations is structurally stable (or "coarse", in the terminology of Andronov et al., 1966, pp. 374 ff.; see also Krasovskiĭ, 1963, ch. 4) if slight changes in its coefficients do not change its stability properties.

Structural stability is generally considered to be an asset for any economic model[27], and seems consistent with a reasonable description of reality: in fact, except in extraordinary circumstances, most actual economic systems appear to be structurally stable even under non infinitesimal changes of the relevant parameters.

Now, the availability of the partial derivatives $\partial \mu_j / \partial \theta_i$ enables one to check structural stability straightforwardly, at least at the local level. The same partial derivatives enable one to determine possible bifurcations, namely the values of the parameters at which a qualitative change in the nature of equilibrium occurs. In fact, to a first approximation, $d\mu_j = \Sigma_i (\partial \mu_j / \partial \theta_i) d\theta_i$ or, if one wishes to consider a particular critical parameter only, $d\mu_j = (\partial \mu_j / \partial \theta_i) d\theta_i$. Therefore, if $\mu_j \neq 0$, letting $\mu_j + d\mu_j = 0$, one can determine the corresponding $d\theta_i = -\mu_j (\partial \mu_j / \partial \theta_i)^{-1}$ (and so the neighbourhood of the bifurcation value of the i-th parameter, $\theta_i + d\theta_i$). In synergetics (Haken, 1978, 1983 ; Medio, 1983; Silverberg, 1983) it is often assumed than when certain critical parameters are changed and the system moves from stability to instability, only a few characteristic roots become positive in real part. By means of sensitivity analysis one can find out whether such parameters exist (instead of simply assuming that they do) and establish which they are.

As regards policy analysis, if policy functions are present in the model, it is possible to ascertain the effect of a change in any one of the parameters in these functions on the dynamic properties of the model in a general way, without having to use numerical simulations.

---

[27] According to some writers, however, its absence should not be automatically considered as a "falsification" of the model: see Vercelli (1982).

This could help with a) the assignment of instruments to targets, if it turns our that some policy parameters crucially affect the stability of the model; and b) the computation of the change that one or more policy parameters must undergo in order that certain characteristic roots, and therefore the dynamic behaviour of the model, are modified in a given direction (for example, to eliminate or reduce instability or an undesired cyclical component).

Furthermore, it may turn out that some parameters related to non-policy variables crucially affect stability. This suggests that the policy authorities should carefully watch the behaviour of such variables and intervene with offsetting actions if the parameters concerned change and approach their bifurcation values.

CHAPTER 3

EMPIRICAL RESULTS

3.1 Parameter Estimates

The model was linearized in the logarithms about sample means in its non-linear parts — eqs. (1), (16), (17) and (21) —. The continuous log-linear model was then reduced to a stochastically equivalent discrete analogue according to the procedure explained in Gandolfo (1981, ch. 3, §3.3.2) which was estimated by using the FIML program "RESIMUL" developed by Clifford Wymer (sample period: 1960-I to 1981-IV).

Some of the parameters included in the model were restricted because the value of the parameter obtained in earlier stages of estimation did not differ significantly from the value to which it was eventually constrained. All parameter restrictions, five in all, increased the efficiency of the estimates[1].

On the whole, estimation results are satisfactory[2]: of 63 estimated parameters, 56 (i.e. 89%) are significantly different from zero at last at the 5% level on asymptotic tests and 51 are also significant at the 1% level. Only two have an incorrect sign (α_{15} and β_{19}), but they are not significantly different from zero[3].

The Carter-Nagar system R^2_W statistics (Carter and Nagar, 1977) is

---

[1] The model is heavily overidentified, and the likelihood ratio test should lead to reject the hypothesis that the over-identifying restrictions as a whole are consistent with the sample (the associated x^2 is 2170 with 505 df). However, since this test is valid only asymptotically and may be invalid for samples as small as ours, we decided to keep the restrictions (a similar decision was adopted in the RBA 1977 model; the other existing continuous time models do not give tests on overidentifying restrictions).

[2] They also stand above the average results so far obtained in the estimation of continuous macrodynamic models. The reader is referred to Gandolfo (1981, p. 210) for information on the average results obtained by the estimated continuous time economy-wide macroeconometric models as known to us. Consideration of subsequent contributions (see, for example, Chiesa, 1979; Jonson and Trevor, 1981; Tullio, 1981; Knight and Mathieson, 1983) does not significantly change those averages.

[3] Note that the ratio of the parameter estimate to its asymptotic standard error

0.568 and the associated χ^2 value is 1656 with 65 df: therefore the hypothesis that the model as a whole is not consistent with the data is rejected.

Let us now discuss the estimation results. We begin by looking at the adjustment parameters (see table 4).

3.1.1 Adjustment Parameters

The parameters α_1, α_3, α_5, α_7, α_9, α_{11}, α_{13}, α_{14}, α_{16}, α_{17}, α_{18}, α_{20}, α_{21}, α_{22}, α' show the speed at which the relevant variable adjusts to its partial equilibrium (target or desired) level. In this case the reciprocal of the adjustment speed can be interpreted as the mean time-lag, i.e. the time needed to eliminate about 63 per cent of the divergence between actual and partial equilibrium values (see Gandolfo, 1981, pp. 13-14); mean time—lags and their standard errors are also shown in table 4.

Table 4 - Estimated adjustment parameters

Parameter	Entering equation number	Point estimate	Asymptotic standard error	Mean time lag (quarters)	Standard error of mean time lag
α_1	(1)	1.423	0.188	0.702	0.093
α_2	(1)	0.151	0.050		
α_3	(2)	1.334	0.220	0.750	0.123
α_4	(2)	0.122	0.011		
α_5	(4)	1.409	0.196	0.710	0.099
α_6	(4)	0.556	0.142		
α_7	(5)	0.673	0.124	1.486	0.275
α_8	(5)	0.902	0.153		

cannot be interpreted as the usual Student's t. With sufficiently large samples this ratio tends to be normally distributed and therefore it is significantly different from zero at the five per cent level ("significant") if it lies outside the interval ±1.96 and significantly different from zero at the one per cent level ("highly significant") if it lies outside the interval ±2.58.

α_9	(6)	1.932	0.287	0.518	0.077
α_{10}	(6)	0.346	0.070		
α_{11}	(7)	0.145	0.056	6.897	2.694
α_{12}	(7)	0.211	0.071		
α_{13}	(8)	0.208	0.059	4.808	1.370
α_{14}	(9)	1.365	0.281	0.733	0.151
α_{15}	(10)	-0.007	0.045		
α_{16}	(10)	0.076	0.019	13.158	3.289
α_{17}	(11)	0.111	0.019	9.009	1.548
α_{18}	(12)	0.286	0.029	3.497	0.354
α_{19}	(12)	1.356	0.062		
α_{20}	(13)	3.445	0.604	0.290	0.051
α_{21}	(14)	0.181	0.079	5.525	2.402
α_{22}	(15)	0.281	0.070	3.559	0.892
α'	(2)	0.050	0.003	20.0	1.213
η	(3)	0.156	0.064		

The highest adjustment speed (i.e. the lowest mean time-lag) is α_{20}, the monetary authorities' adjustment speed to the desired value of m, the proportional rate of growth of the stock of money. This result is not surprising since it is reasonable to expect that monetary authorities are able to adjust the value of their instrument quite rapidly. In our case the mean time-lag is less than a month.

The lowest adjustment speed is α': in fact, the discrepancy between the actual and desired capital stock — which determines the desired rate of capital formation — is adjusted with a mean time-lag of twenty quarters. On the contrary α_3 — i.e. the adjustment speed of the rate of change in the capital stock to its desired value — is quite high, the mean time-lag being just over two months.

The rate of change of k is also strongly influenced by the rate of change of a; this confirms the important role of credit expansion in favouring capital accumulation in the Italian economy.

The point estimate of η, the adaptive expectations coefficient, is low (0.156) but significant, which means that operators are rather slow in changing output expectations[4].

If one considers the results concerning η, α_3, α' and α_4 one finds that firms take a long time both in changing their views about the future and in adapting their desired fixed capital stock to them, but take a much shorter time in changing the speed at which accumulation proceeds. This speed is also influenced by the state of credit conditions.

Real consumption shows a mean time-lag of just over two months, which is rather a short time. This seems a sensible result especially in a period of high inflation — such as the one covered by the second half of our sample — when households are prompted to anticipate rather than to postpone the implementation of their consumption plans.

In addition to their target value, real imports adjust also to a discrepancy between the actual and desired stock of inventories. The "own" adjustment speed of real imports (α_5) is almost three times as high as the second adjustment parameter (α_6). This result seems to support the hypothesis that different behavioural assumptions should be made about these two components of import performance[5].

The fact that real inventories take a rather long time to adjust to their target value is confirmed by the estimate of α_{10}, the adjustment parameter of inventories included in the output equation.

Real exports show a mean time lag of over four months ($1/\alpha_7=1.486$) while the high, and highly significant, value of α_8 is consistent with

---

[4] In principle, in order to estimate η, observations on $\tilde{Y}$ are needed and these, in turn, can be produced only knowing η. However, by means of suitable manipulations (fully described in appendix I, §I.3), $\tilde{Y}$ was eliminated wherever it appeared. This raised the order of the equations which contained $\tilde{Y}$ and so a new endogenous variable ($y \equiv D\log Y$) was defined and introduced for estimation purposes. The model was thus maintained in first-order form; this procedure allowed the estimation of η simultaneously with the remaining parameters.

An alternative, simpler, but much less accurate and much more time consuming procedure was chosen by the authors (see Gandolfo, 1981, and Gandolfo and Padoan, 1980, 1981a,b,c, 1982a,b) in the previous versions of the model, following Knight and Mathieson (1979); the major shortcoming of this procedure is that it did not allow endogenous estimation of η.

[5] We tested the hypothesis of a significant difference between α_5 and α_6; the results are as follows: $\alpha_5-\alpha_6=0.853$ with an a.s.e. of 0.190.

the hypothesis that whenever a credit squeeze occurs (i.e. when the rate of increase of the stock of bank advances slows down) exporting firms increase their selling efforts in order to cope with their liquidity problems as explained in §2.2.1; it should be remembered that α_8 enters with a negative sign in eq. (5). This also lends support to the hypothesis that whenever domestic demand falls (insofar as this is the consequence of a credit restraint) Italian firms try to increase foreign markets penetration.

The value of α_9 indicates that producers are fairly quick (the mean time lag is about one and a half months) to adjust real output to its expected value, while, as we said above, expectations are revised quite slowly.

The adjustment speed of the price of output is fairly low ($\alpha_{11}=0.145$) although quite significant. The fact that prices show a mean time-lag of over a year and a half means that in situations of high and persistent inflation any inflationary impact (which raises $\hat{P}$) produces a lasting effect, since it takes a long time to be absorbed fully by the economy. Our result is consistent with the findings of Milana (1980) who has shown — in a simulation study of the impact of oil price increases on some industrial countries, carried out by using input-output techniques — that unlike, say, Germany, in the case of Italy a long time is needed for the impact of oil price increases to spread out fully through the system. This also could account — in part at least — for the fact that Italy is an "inflation prone" country (to use Corden's (1977) terminology).

The value of α_{12} is highly significant and this supports the idea that — *ceteris paribus* — a monetary expansion (increase in the rate of growth of the money stock) has an inflationary impact on domestic prices. We wish to stress however that such a result need not be interpreted along monetarist lines. As we discussed in the previous chapter, the hypothesis embedded in eq.(7) is that the desired price level is formed according to a mark-up mechanism; monetary expansion may (and in fact does) help price makers to achieve their price targets[6]. Monetary expansion, in other words, does not directly influence prices through an excess demand effect.

The adjustment speed of export prices is also low but higher than that of domestic prices and this indicates that inflationary pressures are absorbed more rapidly in international markets. Real wages, on the contrary, show a mean time-lag of about two and a half months, i.e. they adjust fairly rapidly to domestic inflationary pressures. This result is also not surprising if we recall that our sample includes a period in which full escalator clauses were introduced in Italy (1975 onwards).

Of the two adjustment parameters appearing in the interest rate equation, α_{15} and α_{16}, the former displays the wrong sign, but is also not significantly different from zero, while the latter is low (as a matter of fact, it is the second lowest adjustment speed) but highly significant. These results suggest that the Italian interest rate is not significantly affected by domestic money market conditions while it is heavily dependent on foreign interest rate behaviour although this influence takes a long time to make itself felt. This means that if we assume that $i_{TIT} = \gamma_{10} i_f$, i.e. that the target rate of the monetary authorities is a function of the foreign rate, then the monetary authorities are quite slow in reaching their target; this seems in conflict with other results of the model, namely that the central bank is rather fast in achieving the money growth target (see above). This finding however has good reasons behind it. First of all — as we mentioned in the previous chapter — eq. (10) represents a "compromise" as it includes both market and policy components and therefore market forces might have a different opinion about the determinants of the target rate.

Secondly, the data chosen for i_{TIT} (long term government bond yield)[7] was forced to support the "catch-all" nature of the interest rate in our model (as a matter of fact i_{TIT} was intended to represent the whole interest rate structure of the Italian economy). Finally,

---

[6] In fact eq. (7) is consistent with the demand for and supply of inflation approach: see Gordon (1975).

[7] This choice is justified by the fact that among the (quite few) series of interest rate data covering the full sample period the one chosen is the best "representative" rate. This series was also used by Herring and Marston (1977) in their analysis of Italian monetary policy.

in the course of the sample period the Bank of Italy pegged the interest rate (from 1966-II to 1969-II) in order to stabilize the domestic bond market (see Fazio, 1979)[8] and this means that the target rate aimed at by the monetary authorities is not just a function of the foreign rate.

The excess demand for money, on the contrary, shows a highly significant influence on consumption ($\alpha_2 = 0.151$ with an a.s.e. of 0.050) supporting both the idea that the speed at which C adjusts to $\hat{C}$ is not a constant and the idea that a disequilibrium in the money market has a spillover effect on real consumption.

The speed at which banks adjust the stock of credit to its target value is quite low and highly significant; the mean time-lag is over two years. This is not surprising since eq. (11) describes the adjustment of a stock[9].

Such an impression is confirmed by the value of α_{18}, the adjustment speed of the stock of net foreign assets, which shows a mean time-lag of almost a year. As one might expect, credit expansion produces a considerable increase in capital outflows.

This last result enables us to draw a first conclusion about the overall behaviour of the model. Monetary policy has a strong influence on the workings of the economy, both directly via the effects of m (and hence M) and indirectly, via the effect on bank advances — see the further discussion of eq. (11) below[10].

As one might expect a monetary expansion (squeeze) improves (de-

---

[8] To take account of this an additive dummy variable, assuming the value 1.0 in 1966-II to 1969-II and zero elsewhere, was introduced in eq. (10). Estimation results, however, did not improve dramatically and so we preferred to discard this solution (also given the fact that a multiplicative rather than an additive dummy variable should have been used; but this would have required non linear estimation methods).

[9] The hypothesis that eq. (11) could be specified as a two-stage adjustment process, as in the case of eq. (2), was tested in a previous and provisional version of the model (Gandolfo and Padoan 1983a, 1983b).

However, estimation results and sensitivity analysis suggested that the two-stage decision hypothesis was not suited for the bank advances equation. We therefore returned to the more traditional formulation.

[10] We must recall that direct credit controls have been introduced by the monetary authorities in the period under consideration (see Fazio, 1979); however, the reasonably satisfactory estimation results suggest that our hypothesis about bank be-

presses) real growth (via α_2 and α_4) but increases (contains) inflation (via α_{12}) and worsens (improves) the balance of payments (via α_8 and α_{19}). This is, of course, only the immediate impact but it is consistent with the view that monetary and credit policy heavily influences the behaviour of the Italian economy.

Both nominal taxes and real public consumption show a rather low adjustment speed. As a matter of fact α_{21} has a value very close to that of α_{11}, the adjustment speed of actual to desired prices, and this is consistent with the fact that nominal taxes closely follow domestic inflation. Public consumption expenditure shows a mean time-lag of almost ten months (we may consider α_{22} to be a mean time-lag if we make the plausible assumption that $\hat{G} = \gamma_{15} Y$ i.e. that the "desired" value of public consumption is a fraction of real output) which is also a reasonable result for the Italian economy, i.e. public expenditure (and this is particularly true of its current component G) lags behind with a considerable pro-cyclical effect.

We now note that the estimates of the adjustment speeds allow a ranking of the variables between "fast motion" and "slow motion" ones. Also note that such a ranking can be tested by using the statistical tests on the significance of a difference; for instance we can test the hypothesis that $\alpha_j > \alpha_i$ by testing whether $\alpha_j - \alpha_i$ is significantly greater than zero (this was carried out above to test the difference between α_5 and α_6).

If one wishes to refer to the approach of synergetics discussed by various authors (see above, ch. 1, §1.2 point 5, and ch. 2, §2.3) one could easily determine the "order" variables and the "slaved" ones.

Our empirical results show that the real and financial *stocks* have the lowest adjustment speeds (highest mean time-lags). This would suggest that real and financial accumulation variables "slave" the remaining ones. A result which is certainly realistic and consistent with the medium term approach followed in our model. Another interesting result, in this respect, is that prices, both P and $PXGS$, should be considered

haviour is fairly well described by eq. (11) even without the explicit consideration of such institutional changes.

Table 5 — Estimates of adjustment lags of financial and real variables obtained from continuous time models(a)

Country	Model	Domestic prices	Domestic output	Exports	Imports	Capital flows
UK	Knight and Wymer (1978)	2.0	0.46	0.70	0.74	40
UK	Jonson (1976)	-	14.28	14.28	8.69	60
Australia	RBA '79 (Jonson and Trevor, 1981)	1.8	0.46	3.40	0.40	5.50
Italy	Gandolfo and Padoan, present model	6.89	0.51	1.48	0.71	3.50

[a] The results for UK and Australia are taken from Horne (1983, table 2). The results for Italy are derived from the estimates of α_{11}, α_9, α_7, α_5, α_{18} respectively (see table 4 above). Also note that our results are not qualitatively different from those obtained in two previous versions of our model (Gandolfo and Padoan, 1982, 1983).

"order" variables. With regard to the possible "order" nature of the rate of interest (which one could infer from the low value of α_{16}) one should be even more cautions for the reasons discussed above. However given the still tentative nature of the approach of synergetics in economics, we do not wish to pursue this matter further, but we should like to point out to those who take this approach that the distinction between "slaved" and "order" variables can be rigorously made by means of empirical estimates instead of *a priori* assumption only.

As regards the asset markets clearing assumption, our results are in line with those obtained from other disequilibrium models in continuous time and surveyed in Horne (1983). As she observes (pp. 94-5) a general finding from these studies (see our table 5) is to find longer lags of capital flows[11] adjustment relative to the adjustment speeds of the trade account and the goods market; these results cast some doubt on the assumption of continuous asset market equilibrium within the observation period and the assumption of perfect capital mobility[12]. The obvious conclusion is that the above assumptions have not general validity and that, in any case, they should be subjected to careful scrutiny and empirical testing before going on building models based on their *a priori* assumed validity. A final consideration: those who take the asset market approach seem to believe, more or less implicitly, that these markets are more "important" than the goods markets. Now, the approach of synergetics indicates that the "order" variables are, in some sense, more fundamental than the "slaved" ones; if this is so, the greater "importance" of the financial stocks would derive from their being slow motion rather than rapid motion variables.

Let us now turn to the discussion of the remaining parameters of the model; these are given in table 6.

3.1.2 Elasticities and Other Parameters

We first note that in eq. (1.1) of the model the disposable income expression $(Y-T/P)$ had to be linearized in terms of the logarithms of output and real taxes; however, because of difficulties in estimating the parameters of the resulting equation, the following approximation was used[13]. Since — under the assumption that β_{24} is not significantly different from one[14] — the partial equilibrium level of real taxes

Table 6 - Other estimated parameters

Parameter	Entering equation number	Point estimate	Asymptotic standard error
β_1	(1)	-0.813	0.168
β_2	(2)	0.0*	
β_3	(1),(10)	0.104	0.198
β_4	(1),(10)	0.864	0.145
β_5	(1),(10)	2.497	0.289
β_6	(4)	0.505	0.105
β_7	(4)	0.507	0.078
β_8	(4)	1.257	0.102
β_9	(5)	0.004	0.160
β_{10}	(5)	0.643	0.214
β_{11}	(5)	1.225	0.161
β_{12}	(7)	0.392	0.164
β_{13}	(7)	0.474	0.108
β_{14}	(7)	0.0*	
β_{15}	(8)	0.713	0.159
β_{16}	(8)	0.286	0.128
β_{17}	(9)	0.682	0.040
β_{18}	(11)	-0.305	0.040

[11] In the sense of flows arising from the discrepancy between the desired and actual stocks of (net) foreign assets.

[12] A contrary result is obtained by Blundell-Wignall (1984), who, by using a continuous time disequilibrium model, finds empirical support for a portfolio model of the Deutschemark effective rate. It must however be noted that his model is limited to the financial sector (the real variables are exogenous), and so it cannot be used to compare adjustment speeds of real and financial variables.

[13] A similar procedure is followed by Knight and Wymer (1978)

[14] Estimation results — see table 5 — show that β_{24} is not significantly different from one at the 5% level.

β_{19}	(12)	-0.116	0.150
β_{20}	(12)	0.169	0.056
β_{21}	(12)	1.051	0.120
β_{22}	(12)	0.940	0.113
β_{23}	(12)	2.694	0.436
β_{24}	(14)	0.871	0.066
λ_5	(9)	0.019	0.001
δ_1	(13)	0.0*	
δ_2	(13)	0.077	0.026
δ_3	(13)	0.275	0.039
δ_4	(13)	0.169	0.019
$\log\gamma_1'$	(1)	-0.316	0.018
$\log\gamma_2$	(1),(10)	-18.935	2.703
$\log\gamma_3$	(2)	1.437	0.015
$\log\gamma_4$	(4)	2.987	0.807
$\log\gamma_5$	(4),(6)	0.115	0.197
$\log\gamma_6$	(5)	-0.405	1.161
$\log\gamma_7$	(7)	0.351	0.012
$\log\gamma_8$	(8)	-0.091	0.028
$\log\gamma_9$	(9)	-0.324	0.091
$\log\gamma_{10}$	(10)	0.0*	
$\log\gamma_{11}$	(11)	-3.967	0.093
$\log\gamma_{12}$	(12)	8.002	0.913
$\log\gamma_{13}$	(13)	0.0*	
$\log\gamma_{14}$	(14)	-0.319	0.529
$\log\gamma_{15}$	(15)	-1.439	0.033

*value imposed

from eq. (14) is

$$\frac{T}{P} = \gamma_{14} e^{-\rho_{12}/\alpha_{21}} Y,\qquad(14')$$

where ρ_{12} is the steady-state rate of growth of T (as given in table 3), desired consumption in eq. (1.1) may be approximated by

$$\hat{C} = \gamma'_1 e^{\beta_1 D\log Y} \left(\frac{P}{PMGS}\right)^{\beta_2} Y,\qquad(1.1')$$

where

$$\gamma'_1 = \gamma_1(1-\gamma'_{14}),\qquad \gamma'_{14} = e^{-\rho_{12}/\alpha_{21}} \gamma_{14}.\qquad(1.1'')$$

Eq. (1.1') was used in estimation, and it is the estimate of $\log\gamma'_1$ rather than $\log\gamma_1$ that is given in table 6. However, by using eq. (1.1") it is possible to determine $\gamma_1=0.723$ (a.s.e.=0.131). It also turns out that γ'_1 and γ_1 are not significantly different (the difference $\gamma'_1-\gamma_1$ is equal to 0.006 with an a.s.e. of 0.039).

The rate of growth of real output has a strong and highly significant inverse effect on desired consumption; i.e. aggregate consumption behaviour is "anti-cyclically oriented". The parameter associated with the Laursen and Metzler effect (β_2) was constrained to zero since earlier estimation showed that this parameter did not significantly differ from that value. The theoretical ambiguity concerning the expected sign (discussed in ch. 2, §2.2.1) has therefore not been eliminated by our empirical results.

The parameters associated with the demand for money appear both in eq. (1) and in eq. (10). The elasticity with respect to i_{TIT} is negative (recall that β_3 enters with a negative sign in M^d) but is not significantly different from zero. The absence of an interest rate structure in our model may be taken as an explanation for this result[15].

---

[15] Caranza, Micossi and Villani (1983) obtain significant estimates of interest rate elasticities in their estimation of the demand for money in Italy assuming a full structure of interest rate differentials.

The elasticities with respect to price and real output are both highly significant and quite different from one another; β_4 is not significantly different from one, suggesting *the absence of money illusion*, so that the demand for money in real terms (M^d/P) does not depend on the price level; β_5 is significantly larger than one. This last result is in line with other recent econometric estimates of the demand for money in Italy (Caranza, Micossi and Villani, 1983) and is also plausible since money-stock/output ratios have shown and increasing trend over the sample period for different definitions of the money supply. In this respect our estimate of β_5 is more satisfactory than that of the corresponding parameter in Caranza, Micossi and Villani (1983), since it is larger (as it should be, also according to these authors).

The parameters appearing in eq. (2) have already been discussed with the exception of γ_3, which is highly significant and has a value of just over four ($\gamma_3 \equiv e^{\log \gamma_3} = 4.211$ with an a.s.e. of 0.315); since this parameter is the capital/output ratio times the reciprocal of the desired degree of capacity utilization ($\gamma_3 = \kappa/u$) its value is reasonable. Since the estimated value of η was already discussed above we may turn to eq. (4).

Desired real imports are significantly influenced by both domestic and import prices. As a matter of fact the estimates of β_6 and β_7 are not significantly different[16] so that the possibility of money illusion indicated in previous studies (see §2.2.1) seems to be rejected. Real income elasticity is greater than one and highly significant.

Kreinin and Warner (1983) have estimated import demand equations assuming different elasticities for domestic and foreign prices for nineteen industrial countries. The results they obtain for Italy are significant and the absolute values of both price elasticities are larger than one when they consider a sample period up to 1970-IV (i.e. excluding both the managed float and oil price shocks). Their results for Italy are not so satisfactory, however, if the period following 1972-I is covered by their sample, with and without oil included

---

[16] $\beta_6 - \beta_7 = 0.001$ with an a.s.e. of 0.052.

in the dependent variable. Since our observations include oil imports and cover the full managed float period, we may consider our estimates to be fairly satisfactory even if the absolute values of the elasticities are lower in our case. The homogeneity tests they carry out suggest that the use of the price ratio is not justified for Italy both for the pre- and the post-1970-IV sample if total imports (including oil) are considered. This last result is not in line with ours as we have shown that β_5 and β_6 are not signficantly different (see note 16 above).

As far as the second component of eq. (4) is concerned, i.e. the adjustment of actual to desired inventories, results show that this component significantly affects real imports behaviour. However, we were not able to obtain a significant estimate of γ_5, the ratio of desired inventories to expected output.

This is an outcome which we also experienced in previous versions of our model[17] and which we may impute to the poor quality of the data concerning inventories.

The parameters entering into $\widehat{XGS}$, the desired level of real exports, are all highly significant with the exception of β_9, the relative price elasticity. This result may be due to the hybrid nature of our equation, which contains both supply and demand elements. Actually several alternative specifications of the price component were tested, including separate elasticities for PF and $PXGS$ or for PF_f, E, and $PXGS$ (where PF_f is the foreign price in foreign currency and E is the exchange rate considered as a separate exogenous variable) as suggested by Kreinin and Warner (1983), but like these authors, we were not successful in finding significant price elasticities. We also tested a more rigorous specification of supply and demand factors in the export equation which was the following, as suggested by Drollas (1976):

---

[17] In the previous version of this model (Gandolfo and Padoan, 1983a and 1983b), γ_5 was constrained to a given value. We chose not to follow this solution in the present version since it did not improve the overall estimation results.

$$D\log XGS = \alpha_7 \log \frac{\widehat{XGS}_D}{XGS} + \alpha'_7 \log \frac{\widehat{XGS}_S}{XGS},$$

where $\widehat{XGS}_D = (\frac{PXGS}{PF})^{-\beta_9} Y_F^{\beta_{10}}$ is the "desired" demand for exports and

$\widehat{XGS}_S = (\frac{PXGS}{P})^{\beta'_9} Y^{\beta_{11}}$ is the "desired" supply of exports.

This version was tested both with the remaining equations of the model as in the original version and with the export price equation specified as in Drollas (1976):

$$D\log PXGS = \alpha_{13} \log \frac{\widehat{XGS}_D}{XGS} + \alpha'_{13} \log \frac{\widehat{XGS}_S}{XGS}.$$

In all these cases the estimates of the export price elasticities were poor, if not insignificant; further, this specification, often considerably affected negatively the estimates of other parameters of the model[18]. Therefore we decided to retain our original formulation. Note that Drollas (1976) also obtained non significant estimates of price elasticities for the Italian export equation.

A rigorous treatment of the combined effect of supply and demand factors on export behaviour for Italy has been carried out by Petit (1981). She estimates with simultaneous methods a non-linear model of two differential equations for XGS and $PXGS$ in which the effect of supply and demand factors is determined via the introduction of a "switching function". This procedure, suggested by Wymer (1979, 1983), allows for an *endogenous* separation of observations according to whether supply or demand factors determine the observed variable (rather than using the standard "ad-hoc" pre-estimation separation of the sample). The encouraging estimation results obtained by Petit suggest that this is the approach that should be followed. This however requires non linear estimation techniques which, at present, are rather problematic when larger continuous models, such as ours, are involved.

Let us now pass on to the remaining parameters of the export equa-

---

[18]Additional constraints on β_9 were also tested. These included constraints derived from steady-state conditions and cases in which β_9 was given an *a priori* value. In all cases results were poor.

tion. The elasticity with respect to world demand (β_{10}) is highly significant and so is β_{11}, suggesting that supply effects are quite important in determining Italian export behaviour.

The parameters appearing in the output equation have already been discussed so let us turn to eq. (7) which pertains to the price of output. Both import prices and nominal wages significantly affect the desired level of the price of output; β_{14}, the elasticity with respect to PROD (exogenous productivity), was constrained to zero since in earlier stages of estimation the values obtained did not differ significantly from this value. This result can be explained by the fact that PROD is defined as industrial value added divided by industrial employment and is therefore not well suited to enter an equation which refers to the price deflator of the total domestic product; alternative definitions of PROD, on the other hand, would have been less suited to define the standard effect of productivity on price formation.

Although monetary expansion does play an important role in "accommodating" Italian inflation, empirical results support the mark-up hypothesis embedded in eq. (7). In fact, the estimate of $\log \gamma_7$ is highly significant and so is that of γ_7 ($\gamma_7 \equiv e^{\log \gamma_7} = 1.420$ with an a.s.e. of 0.017) and yields a reasonable value of the mark-up factor.

Both domestic and foreign prices have a significant influence on the desired level of export prices. Estimation results suggest that the effect of P is much larger than that of PF. This means that although foreign competitive pressures are important in determining the pricing behaviour of the Italian exporting firms, domestic inflation is much more important.

The desired value of money wages is significantly affected by the level of the price of output and also shows a highly significant trend consistent with the hypothesis that exogenous factors ("social pressure") influence nominal wage performance; empirical results confirm that the wage-price spiral which is present in our model is — *ceteris paribus* — stable [19]. All the parameters of eq. (10) have been discussed

---

[19] As regards distribution, parameter estimates suggest that in the steady-state it will shift in favour of wage-earners, for the expression $(\beta_{17}-1)\lambda_1 - \lambda_4 + \lambda_5$ discussed in §2.3 is significantly positive (it equals 0.013 with a.s.e. 0.001). As regards

above with the exception of γ_{10} which was constrained to 1.0.

The next equation to be discussed is eq. (11) which explains the behaviour of bank advances. The desired stock of bank advances is significantly related to the level of the domestic interest rate. The inverse relation shows that acquisition of government bonds represents an alternative investment opportunity for banking firms, a typical feature of the Italian financial system. The results of this equation make it possible to appreciate the effects of monetary restrictions on the behaviour of the economy more fully. The importance of credit expansion on real and financial accumulation (via α_4 and α_{19}) as well as on export performance (via α_8) has already been stressed. A situation of monetary tightness negatively affects the expansion of bank advances through two different channels. If international monetary conditions become more strict and if this is reflected in a higher level of foreign interest rates, the domestic rate rises accordingly — see eq. (10) —. This shifts bank investment preferences from advances to government bonds[20].

If the domestic money supply rises at a lower rate — see eq. (13) — the desired level of bank advances also grows more slowly given the value of the interest rate and of γ_{11}, which includes the relationship between the stock of credit and the monetary base.

Of the five elasticities included in $\widehat{NFA}$, the desired stock of net foreign assets, four are highly significant. The domestic interest rate stability, consider the subsystem

$$D\log P = \alpha_{11}[\log W^{\beta_{13}} - \log P], \qquad D\log W = \alpha_{14}[\log P^{\beta_{17}} - \log W],$$

which has the characteristic equation $\mu^2 + (\alpha_{11} + \alpha_{14})\mu + \alpha_{11}\alpha_{14}(1-\beta_{13}\beta_{17}) = 0$. The necessary and sufficient condition for this sub-system to be stable is $1 - \beta_{13}\beta_{17} > 0$; this condition is fulfilled since $1 - \beta_{13}\beta_{17} = 0.677$ with an a.s.e. of 0.074.

[20] One has to recall, however, that for quite a number of the years covered by our sample, Italian banks have been obliged by the monetary authorities to invest in predetermined amounts of government bonds. Therefore they were not entirely free to choose between alternative investment opportunities.

does not seem to affect Italian capital flows, while a significant effect, though not high in absolute value, is displayed by the foreign rate. The parameter β_{23} is positive and highly significant; this allows us to assume that Q, the ratio of the forward to the spot exchange rate, may be considered as a proxy for exchange rate expectations, and so whenever Q rises, expected depreciation increases capital outflows. This also means (see ch. 2) that the interest rates included in $\widehat{NFA}$ may be considered as proxies for short-term interest rates[21]. Nominal domestic and foreign output are significant scale variables; this also means that whenever the Italian economy displays a higher nominal growth (which may also be the result of higher inflation and lower real growth, i.e. higher stagflation), the Italian balance of payments worsens both in the current account (higher imports and lower exports) and in the capital account.

Estimation results of the central bank's reaction function well reflect the problems discussed in ch. 2 concerning the institutional difficulties which the Italian monetary authorities have to face in pursuing their targets; the parameters associated with money creation originated by the government and the balance of payments, δ_3 and δ_4 respectively, are both highly significant while only one of the two δ's included in the "true" objective function of the Bank of Italy — δ_2, the weight assigned to the inflation differential[22] — turned out to be significantly different from zero and with the expected sign. This result seems to conflict both with our previous results in earlier versions of the model and with recent econometric research on monetary policy in European countries (Demopoulos, Katsimbris and Miller, 1983).

Earlier versions of our model (Gandolfo and Padoan, 1982a, 1983a and 1983b) included a desired rate of growth of output among the arguments of the monetary authorities' reaction function in addition to those

---

[21] This may also account for the fact that the parameter associated with i_{TIT} is not significantly different from zero.

[22] The parameter associated with the desired reserves/import ratio was eventually constrained to zero (and, therefore, $\log \gamma_{13}$ was constrained to zero as well) because in earlier stages of estimation we could not obtain significant estimates. It should be noted that this ratio was also insignificant in Black's (1983) analysis, who esti-

included in the present version, while Demopoulos, Katsimbris and Miller (1983) estimated a central bank's reaction function on very similar lines for Italy[23]. In both cases the number of significant parameters is higher than in the present version. In both cases however the theoretical specification is unsatisfactory, since it does not provide a rigorous separation between the components which are directly controllable and those which are not — in the sense specified in ch. 2 — by the central bank. Therefore it would seem that, once this separation is clearly made, the Bank of Italy is able to pursue only one independent target, given the "constraint" imposed by the other two channels of monetary creation.

One should add, however, that a shortcoming of the present specification of the central bank's reaction function is that, although our sample covers both the fixed exchange-rate and the managed float period, we did not allow for the change in the form of the reaction function according to the change in the state of the economy. This would require the "switching function" method (Wymer, 1979, 1983) and non linear estimation.

The discussion of the estimation results concludes with eqs. (14) and (15). The elasticity of nominal taxes with respect to nominal income is highly significant and slightly lower than one (however, as we said asbove, β_{24} is not significantly different from one at the 5% level); as we mentioned in ch. 2, β_{24} should be greater than one to reflect fiscal drag effects, if T included direct taxes only, but since T includes all kinds of taxes — many of which are proportional or less than proportional to nominal income — our result is not surprising. The estimated value of γ_{15} which represents the share of public expenditure in income is rather low ($\gamma_{15} \equiv e^{\log \gamma_{15}} = 0.238$ with an a.s.e. of 0.009) but highly significant. An explanation of this result is that

---

mates a monetary authorities' reaction function with the discount rate as the policy instrument.

[23]The policy variable (money supply) is defined in slightly different terms and single equation estimation is carried out by these authors.

G is defined as public consumption, which is only a part of total public expenditure [24].

3.2 Stability and Sensitivity

The (local) stability of the model, given the parameter estimates, can be examined by using the linear approximation about the steady-state derived in Appendix I, §I.2. Table 7 presents the estimates of the characteristic roots of this linear approximation, together with their asymptotic standard errors. We remind that the damping period is the time required for about 63% of the initial deviation to be eliminated (of course the damping period is defined only for stable roots); the period of the cycle corresponding to the complex roots is defined in the usual way.

Table 7 - Characteristic roots of the model

Root	Asymptotic standard error	Damping period (quarters)	Period of cycle (quarters)
-0.0006	0.0006	1605.136	
-0.0186	0.0021	53.878	
-0.0250	0.0017	40.069	
-0.0286	0.0034	35.005	
-0.0726	0.0211	13.772	
-0.0941	0.0414	10.622	
-0.1337	0.0359	7.477	
-0.1873	0.0895	5.338	

---

[24] The fact that G is not total public expenditure means that $PG - T$ is not the "true" public deficit. Therefore eq. (21), which defines the public sector borrowing requirement as equal to $PG-T$, had to be modified by the introduction of an exogenous adjustment factor in order to ensure data consistency. This shortcoming has its origin in the difference of presentation between national accounting data (from which P, G and T are taken) and public finance data (from which H is taken). For details see appendix II, §II.1.2.

−0.2078	0.0589	4.812	
−0.2743	0.0692	3.645	
−0.8150	0.1397	1.227	
−1.2785	0.2126	0.782	
−1.4076	0.1983	0.710	
−1.4475	0.2031	0.691	
−2.4003	0.2710	0.417	
−3.3892	0.5774	0.295	
−0.0867 ±0.1477i	0.0455, 0.0300	11.538	42.526
−0.4269 ±2.3158i	0.0785, 0.3191	2.343	2.713

All the real characteristic roots are negative[25], and all the complex roots have negative real parts. The steady-state path is locally stable; the convergence, however, will not be monotonic since cycles will occur: a medium-long cycle (having a period of about eleven years) and a short cycle (with a period of about eight months).

Therefore we can conclude that our model gives rise to a *stable cyclical growth* behaviour, which is certainly the type of movement which Italy (and most other industrialized countries) experienced in the period under consideration. It should also be pointed out that, depending on the conditions determining the arbitrary constants in the general solution of the linearized model, in certain cases the cyclical components may also bring about a *decrease in the level* of the relevant variables (real output, for instance). This would not be inconsistent with the most recent past and with the possible future behaviour of some industrialized countries, including our own.

As regards *sensitivity analysis*, its importance has already been stressed in ch. 2, section 2.3. Since there are 20 characteristic roots

---

[25] To be precise, the first real root has an asymptotic confidence interval (at the 5% level) which includes positive values. However, should this root be positive, we can say that it will not be greater than 0.000635 (at the 5% level), a value so small than can be neglected for all practical purposes.

and 63 estimated parameters, it would be inconvenient and uninteresting to print the full 20x63 sensitivity matrix. We therefore decided to present a selection of the results, according to the principles expounded in section 2.3, namely:

a) the partial derivatives of the characteristic roots with respect to all policy parameters;

b) some particularly large (in relative terms) partial derivatives which imply that the parameter concerned crucially affects stability.

These two criteria are only partially overlapping, for not all the partial derivatives selected according to a) would have also been selected according to b); naturally, the partial derivatives included in b) are *additional* with respect to those included in a), in the sense that they do not refer to policy parameters.

Before passing to comment on the results, we would like to point out that our model appears to be *structurally stable*. In fact, examination of the full sensitivity matrix[26] reveals that slight changes in any one of the parameters do not change the stability properties of the model at least locally. This, as was noted in section 2.3, seems consistent with a reasonable description of reality.

Generally speaking, the parameters that can be considered as policy parameters are those included in the monetary authorities' reaction function ($\alpha_{20}, \delta_1, \delta_2, \delta_3, \delta_4, \gamma_{13}$), in the tax revenue equation ($\alpha_{21}, \gamma_{14}, \beta_{24}$), in the public expenditure equation (α_{22}, γ_{15}), and in part of the interest rate equation (α_{16}, γ_{10}).

As regards the monetary authorities' reaction function, its detailed derivation (see ch. 2, §2.2.4) shows that δ_3 and δ_4 are not full policy parameters, in the sense that they can be influenced only to a limited extent by the monetary authorities, but this suffices to include them in category a). On the policy nature of the various parameters present in the tax revenue and in the public expenditure equations no doubt

---

[26] Given the criteria a) and b) above, all the partial derivatives excluded are smaller (in absolute value) than those included, so that to support our statement it is sufficient to look at tables 8 and 9 below.

should exist (at least in principle, but many doubts arise in the Italian context). The possible partial policy nature of α_{16} and γ_{10} has already been discussed in ch. 2, §2.2.3, and in §3.1 above. Finally note that δ_1, γ_{13}, and γ_{10} are not present in table 8 for they were constrained during estimation.

Let us now examine the results given in table 8. For convenience of the reader we have denoted by an asterisk those partial derivatives which also satisfy criterion b).

We first note that δ_3, β_{24}, γ_{14} do not appreciably affect any of the characteristic roots of the model, and this has interesting implications. As regards δ_3, the meaning is that a change (increase or decrease) in the parameter which reflects the effect on money creation of the change in the public sector's borrowing requirement does not *directly* affect the stability of the model. It may, of course, indirectly affect it through the effect on money creation (see below).

The lack of use of taxation as a policy tool has already been discussed in ch. 2, section 2.2.4, so that the absence of effects of β_{24} and γ_{14} on stability is not surprising. The favourable effect of an increase in α_{21} on stability (and precisely on the eight real root) is not inconsistent with this finding, for it simply means that an increase in the speed of tax collection (namely the speed at which actual taxes adjust to their partial equilibrium level) is stabilizing, which is true independently of the fact that this partial equilibrium level is policy determined or not.

An increase in α_{20} (the monetary authorities' adjustment speed of actual to target values of the rate of change of the money supply) is generally stabilizing, especially on the last real root and on the real part of the first pair of complex conjugate roots. A similar result holds for α_{21} (discussed above), for α_{22} (the adjustment speed of actual to target values of public expenditure) and for α_{16} (the adjustment speed of the domestic to the foreign interest rate; if the latter is interpreted as a target rate, α_{16} becomes a "true" adjustment speed). All these results are consistent with the general idea that, when a model is stable, an increase in the adjustment speeds of

Table 8 - Sensitivity analysis with respect to the policy parameters

Root (μ)	$\partial\mu/\partial\alpha_{20}$	$\partial\mu/\partial\delta_2$	$\partial\mu/\partial\delta_3$	$\partial\mu/\partial\delta_4$	$\partial\mu/\partial\alpha_{21}$	$\partial\mu/\partial\beta_{24}$	$\partial\mu/\partial\gamma_{14}$	$\partial\mu/\partial\alpha_{22}$	$\partial\mu/\partial\gamma_{15}$	$\partial\mu/\partial\alpha_{16}$
-0.0006	0.0000	-0.0003	0.0000	0.0000	-0.0000	-0.0000	0.0000	0.0000	-0.0000	-0.0000
-0.0186	-0.0002	-0.0000	-0.0011	0.0060	0.0028	-0.0022	0.0028	-0.0001	0.0005	0.0008
-0.0250	0.0	-0.0000	-0.0000	0.0000	0.0000	0.0000	0.0000	0.0000	-0.0000	-0.0000
-0.0286	0.0005	0.0000	0.0020	-0.0140	0.0008	-0.0001	0.0013	-0.0004	-0.0024	-0.0000
-0.0726	0.0000	0.0000	0.0000	-0.0002	-0.0000	0.0000	0.0001	0.0000	-0.0000	-1.0080*
-0.0941	-0.0000	-0.0001	0.0000	0.0002	0.0004	-0.0001	-0.0001	0.0000	0.0001	0.0070
-0.1337	-0.0001	0.0000	0.0000	0.0010	0.0380	-0.0035	-0.0179	-0.0020	0.0259	0.0007
-0.1873	-0.0000	0.0000	0.0006	-0.0004	-1.1259*	-0.0074	-0.0105	0.0020	0.0009	0.0002
-0.2078	-0.0000	0.0000	0.0000	0.0001	-0.0011	-0.0000	-0.0002	0.0000	0.0002	0.0000
-0.2743	0.0000	-0.0000	-0.0001	0.0001	0.0105	0.0007	0.0010	-0.9770*	0.0268	0.0000
-0.8150	-0.0029	0.0018	-0.0005	0.0460	-0.0637	-0.0049	-0.0607	0.0104	0.0893	0.0004
-1.2785	-0.0013	0.0008	0.0001	0.0161	0.0014	0.0001	0.0016	-0.0000	-0.0031	-0.0000
-1.4076	-0.0011	-0.0064	-0.0002	0.0154	0.0166	0.0015	0.0151	-0.0000	-0.0021	-0.0000
-1.4475	0.0046	-0.0140	-0.0004	-0.0510	-0.0417	-0.0039	-0.0368	0.0000	0.0013	0.0000
-2.4003	0.0186	-0.0370	-0.0005	-0.1023	-0.0140	0.0144	-0.1611	0.0236	0.0831	0.0001
-3.3892	-0.7677*	2.1151*	0.0242	-0.7403	-0.0073	-0.0015	0.0007	-0.0010	-0.0440	-0.0001
-0.0867 ±0.1477i	-0.0759*, 0.0504	-0.0055, -0.0057	0.0008, 0.0010	1.6335*, -0.9900	-0.0454, 0.0258	0.0001, -0.0002	-0.0834, 0.0477	0.0013, -0.0010	0.1263*, -0.0750	-0.0006, -0.0085
-0.4269 ±2.3158i	0.0003, 0.0011	-0.0019, 0.0001	-0.0000, -0.0000	0.0099, -0.0235	0.1030, -0.0990	0.0034, -0.0048	0.1538, -0.1398	-0.0274, -0.0117	-0.1153, -0.0497	0.0000, 0.0000

actual to desired (or partial equilibrium) values is stabilizing, at least up to a certain point[27].

To complete the examination of the effects of changes in the parameters included in the monetary authorities' reaction function we now consider $\partial \mu / \partial \delta_2$ and $\partial \mu / \partial \delta_4$. An increase in δ_2 (the weight attached to the relative inflation term) has a destabilizing effect on the last real root. The value of the partial derivative under consideration (2.1151) may seem huge with respect to the other partial derivatives, but to assess its practical importance we must compare it with the values of the parameter ($\delta_2=0.077$) and of the characteristic root ($\mu_{16}=3.3892$). Assuming that the linear approximation is good enough also for finite changes, the bifurcation value of δ_2 (see ch. 2, §2.3) lies in the neighbourhood of 1.679, which means an increase[28] of 1.602 in absolute terms and of 2081% (almost 21 times) in relative terms, which is far too great to be likely.

As regards δ_4 (the effect on the rate of money creation of a change in the rate of change of international reserves), an increase in this parameter has a destabilizing effect on the real part of the first pair of complex roots; its bifurcation value lies in the neighbourhood of 0.222, which means an increase of 0.053 in absolute terms and of 31.5% in relative terms, and this is a possible value. It follows that the monetary authorities should continue to carefully control[29] the effects of changes in foreign exchange reserves upon the money supply in order to avoid instability (which would take the form of undamped

---

[27] This proviso is due to the fact that, since there are some positive though very small partial derivatives, too great an increase might cause the corresponding roots to become unstable. All this, however, must be considered with the utmost caution, for it involves changes in the parameters far too large for their effects to be examined by means of first differentials only.

[28] Actually, the linear approximation *is* fairly good, for it only slightly "overshoots" the true bifurcation value. In fact, if we substitute $\delta_2=1.679$ in the place of its estimated value and recompute the characteristic roots of the model, we find that μ_{16} becomes slightly positive ($\mu_{16}=0.051$). Similar results hold in the following analysis.

[29] The fact that they did (in the sample period) is shown — apart from historical knowledge — by the stability of the model.

cycles).

It remains to be discussed the effect of γ_{15} (the desired ratio of public expenditure to domestic output). An increase in γ_{15} has a destabilizing effect on the real part of the first pair of complex roots; its bifurcation value lies in the neighbourhood of 0.923^{30}, which means an increase of 0.686 in absolute value and of 290% in relative terms. This value — which would mean that public expenditure absorbs 92% of domestic output — can be ruled out as impossibly high. It remains true, however, that increases in γ_{15} will decrease the damping of the long cycle (i.e. will increase the damping period) and increase its period.

We now turn to the results selected on the basis of criterion b), presented in table 9. Most of them concern adjustment speeds: this is a

Table 9 - Sensitivity analysis with respect to other selected parameters

Root (μ)	$\partial\mu/\partial\alpha_6$	$\partial\mu/\partial\alpha_{10}$	$\partial\mu/\partial\alpha_{12}$	$\partial\mu/\partial\alpha_{17}$	$\partial\mu/\partial\beta_{13}$	$\partial\mu/\partial\beta_{17}$
-0.0006	-0.0000	0.0000	0.0001	0.0000	-0.0000	-0.0000
-0.0186	0.0032	-0.0031	0.0000	-0.0017	0.0002	0.0000
-0.0250	-0.0000	0.0000	-0.0000	0.0000	0.0000	-0.0000
-0.0286	0.0065	-0.0120	-0.0000	0.0050	-0.0011	-0.0000
-0.0726	-0.0000	0.0000	-0.0004	0.0011	-0.0004	-0.0003
-0.0941	0.0001	-0.0001	0.0019	0.0105	0.1036*	0.0719*
-0.1337	0.0785	-0.1083	0.0002	0.0047	-0.0024	-0.0001
-0.1873	0.0215	-0.0260	-0.0020	0.0118	-0.0011	0.0003
-0.2078	0.0003	-0.0003	-0.0011	0.0050	0.0003	0.0002
-0.2743	-0.0159	0.0180	0.0003	-0.0009	0.0015	-0.0000
-0.8150	0.1933	-0.2222	0.0117	0.0534	-0.0091	-0.0006
-1.2785	-0.0020	-0.0020	0.0016	0.0238	-0.0003	-0.0002
-1.4076	-0.0061	0.0020	0.0178	-0.0211	-0.0515	-0.0362
-1.4475	0.0190	-0.0113	-0.1050	0.0992	-0.0718	-0.0470

[30] Table 6 gives the point estimate of $\log \gamma_{15} = -1.439$; the corresponding natural value is $\gamma_{15} = 0.237$ (a.s.e.=0.008). The partial derivatives were computed with respect to γ_{15}.

−2.4003	−0.2741	−0.2821	−0.2538	0.3764	0.1100	−0.0015
−3.3892	−0.0084	0.0401	2.9544*	−5.4117*	−0.0275	0.0087
−0.0867 ±0.1477i	0.0780*, −0.0429	−0.0689, 0.0328	0.0458*, 0.0201	1.9422*, −0.3359	0.0694*, −0.0419	0.0030, −0.0013
−0.4269 ±2.3158i	−0.0372, 2.2186*	0.3371*, 0.7031*	0.0190 0.0017	−0.0200, −0.0472	−0.0454, −0.2215	−0.0003, −0.0014

confirmation of the importance of these parameters. The partial derivatives $\partial\mu/\partial\alpha_6$ and $\partial\mu/\partial\alpha_{10}$ are best considered together, for they represent the effect on stability of the parameters associated with the inventory terms in the import and in the output equations. An increase in these parameters has a destabilizing effect on the complex roots. Therefore, if the reaction of imports and of producers to a discrepancy between the desired and actual level of inventories is too great, unstable *inventory-determined cycles* will occur: note that not only the amplitude will be increasing, but also the frequency will be greater (in general, a greater frequency of a cycle is an undesirable feature). As regards α_6, its bifurcation value (referred to the real part of the first pair of complex roots) lies in the neighbourhood of 1.668, namely an increase of 1.112 in absolute terms and of 200% in relative terms. Such an increase seems unlikely, but note that smaller increases in this parameter, though preserving the stable nature of the roots, increase the frequency of the oscillations of the short cycle. For example, a 65% increase in α_6 — which is a reasonable value, for it falls at the upper limit of the 1% asymptotic confidence interval — would increase the frequency of the short cycle, for its period would pass from about 8 months to 6 months.

As regards α_{10}, an increase of 1.266 in absolute terms (366% in relative terms) would be necessary to reach the neighbourhood of its bifurcation value (referred to the real part of the second pair of complex roots). The increase compatible with the 1% asymptotic confidence interval is 52%, which would not appreciably increase the frequency of the short cycle (its period would pass to about 7.7 months). Therefore we conclude that the greater destabilizing capacity of inventories lies in their influence on imports, a finding consistent with the behaviour of the Italian economy, where inventory accumulation (or decumulation) through imports in the different phases of the cycle plays an important role.

An increase in α_{12} (the parameter which represents the effect on the rate of inflation of the acceleration of money creation) is destabilizing, for it unfavourably affects the last real root and the real part of the first pair of complex roots. This destabilizing effect is economically obvious, but it should be noted that it is not as strong as some schools of thought would suggest. In fact, the bifurcation values of α_{12} lie in the neighbourhood of 1.358 (an increase of 1.147 in absolute terms and of 544% in relative terms) and 2.104 (an increase of 1.893 in absolute, and of 897% in relative, terms) respectively. These are not plausible values, considering also that the upper value of the 1% asymptotic confidence interval for α_{12} is 0.1832 (+87%).

This result is best seen in conjunction with the effects of β_{13} (elasticity of prices with respect to the wage rate) and β_{17} (elasticity of the wage rate with respect to prices), for reasons that will be made clear below. Increases in β_{13} and β_{17} have a destabilizing effect on the sixth real root and (as regards β_{13} only) on the real part of the first pair of complex roots. The corresponding bifurcation values, however, are beyond any reasonable value, for they by far exceed one (these values lie in the neighbourhood of 1.382 and 1.723 as regards β_{13}, and of 1.991 as regards β_{17}). Although we do not wish to enter into the controversy of monetarist versus other explanations of inflation, we would like to make a few general considerations. Our price equation is eclectic, as explained in detail in ch. 2, §2.2.2, and the data have not rejected our hypothesis that mark-up and institutional factors (wage-indexation etc.) on the one hand, and monetary acceleration on the other, play an important role. Therefore "unilateral" explanations (of the kind "monetary factors are unimportant" or "inflation is determined by monetary indiscipline only") cannot be accepted as regards the Italian economy. Also, sensitivity analysis suggests that all the factors considered have destabilizing effects; these, however, do not seem sufficiently strong to destabilize the system *by themselves alone*, of course if one excludes preposterous values of the parameters.

It remains to discuss the partial derivatives $\partial\mu/\partial\alpha_{17}$. This parameter, which represents the adjustment speed of the actual to the desired level in the bank advances equation, has two conflicting effects. An increase in α_{17} has a strongly stabilizing effect on the last real root on the one hand, and a strongly destabilizing effect on the real part of the first pair of complex roots on the other. It can be shown that the latter effect prevails on the former, so that a decrease in α_{17} has a favourable effect. In fact, negative values of α_{17} are impossible, and its bifurcation value (referred to the last real root) lies in the neighbourhood of -0.626. On the contrary, the bifurcation value referred to the real part of the first pair of complex roots lies in the neighbourhood of 0.156 (an increase of 0.045 in absolute, and of 40% in relative, terms), which falls within the 1% asymptotic confidence interval whose upper limit is 0.160.

The interesting conclusion is that too great a value of the adjustment speed in the bank advances equation can give rise to an unstable cyclical behaviour of the model (a sort of *credit-induced cycle*). If we recall that credit can be classified as an "order" variable (see §3.1), this result confirms the crucial role played by credit in our model.

3.3. Predictive Performance

The predictive perfomance of the model can be analyzed either by using the linearized discrete analogue that has been employed for estimation or by using the original non-linear continuous model. For the reasons explained at some length in Gandolfo (1981, ch. 3, section 3.4) we adopted the second alternative, which amounts to substituting the parameter estimates in the differential equation system laid out in table 1[31] and then solving it. For this solution one only needs the initial values of the endogenous variables and the time paths of the exog-

---

[31] As modified to eliminate $\tilde{Y}$: see Appendix I, section I.3.

enous variables, as explained in ch. 1, §1.2, point 8. If we recall the distinction between single-period and dynamic forecasts in continuous models made there [32], it is self-evident that the former involve a procedure which is much more time-consuming and costly (in terms of computer time). But, apart from this, we believe that dynamic forecasts give a much better idea of the forecasting ability of the model, especially if one wants to produce forecasts for more than one period ahead and/or for time intervals different from that inherent in the data.

Therefore, though dynamic forecasts are generally poorer than single period forecasts, we decided to use the former to examine the predictive performance of the model. To this purpose we produced both in-sample and out-of-sample forecasts; the latter were also of the *ex post* type, for we wanted to eliminate the source of error due to inexact values of the exogenous variables. Furthermore, since we did not want to test how good or bad we are at producing forecasts by using a model *and* judgement, but wanted to test the model as such, we did not make any adjustment to it, and so it was used mechanically.

In sum, we used the model for

a) producing in-sample dynamic forecasts (1960-I to 1981-IV);
b) producing out-of-sample dynamic *ex post* forecasts for the period 1982-I to 1983-IV.

The forecasts were produced with reference to the same time interval inherent in the data (the quarter) for obviuous reasons of comparability with the actual values. In order not to burden this work with additional and lengthy tables, we do not give the actual and computed values (these, however, are available from the authors on request) but only the root mean square errors.

Let us begin by a). The root mean square errors[33] are given in table 10. It should be remembered that the variables are expressed in loga-

---

[32] It should be pointed out that, if one employs the linearized discrete analogue, then the distinction between single-period and dynamic forecasts is the same as that used in discrete models.

[33] Given the way in which the "estimated" values are obtained, no standard errors of forecast could be computed.

rithms or as percentages: consequently, the RMSE gives the average error as a proportion of the actual level of the endogenous variable.

Table 10 - In-sample root-mean-square errors
First quarter 1961 - fourth quarter 1981

Variable	Root-Mean-Square Error of Dynamic Forecasts
C	0.016
k	0.003
MGS	0.072
XGS	0.071
Y	0.016
P	0.015
$PXGS$	0.026
W	0.020
i_{TIT}	0.166
A	0.019
NFA	0.087
m	0.009
T	0.031
G	0.056
V	0.203
R	0.251
K	0.005
M	0.002
H	0.157
r	0.173
a	0.019
h	0.050

Seventeen out of 22 variables have errors below 10 per cent and for 13 variables the error is below 5 per cent. It is also encouraging to see that low errors are present in both real and financial variables. This can be considered as a further indications of the satisfactory integration of real and financial aspects in the model.

The highest error (25 per cent) is dispayed by international reserves, but this is not surprising given the volatility of this variable; the same reason explains the large error in the rate of change in international reserves.

A relatively high error is associated also with the stock of inventories. In this case statistical reasons may be advanced, in addition to economic ones, to explain this result. Changes in inventories usually behave in a highly volatile manner in advanced economies; in addition, data for changes in inventories in Italy are defined as a residual after the main components of national accounts have been computed, given the overall balance constraint.

Two other variables display an error which is higher than 10 per cent: the public sector's borrowing requirement, and the interest rate. As far as the former is concerned, is should be noted that, given the non coincidence between national accounts from which P, G and T are taken and public finance data, from which H is taken, eq. (21) had to be modified for estimation purposes (see appendix II, §II.1.2) with the addition of an exogenous correction factor. Hence, in this case too, statistical reasons may be advanced to explain the relatively high error in H. From the economic point of view we may note that this variable has certainly experienced pronounced volatility in the Italian case, especially in recent years.

Estimation results of the interest rate equation have proved to be not completely satisfactory also given the hybrid nature of this function (see §3.1.1). The relatively high error in i_{TIT} confirms these difficulties.

As regards b), the out-of-sample root mean square errors, presented in table 11, are higher than in-sample ones but this is not surprising, both because forecasts are *dynamic* and because the *observed* values for 1983 are provisional.

Nonetheless only 6 out of 22 variables display an error higher than 10 per cent. Also note that the variables with the highest error are the same as those of the in-sample case. The only additional variable which is this case displays an error higher than 10 per cent is *NFA*, but

Table 11 - Out-of-sample root-mean-square errors
First quarter 1982 - fourth quarter 1983

Variable	Root-Mean-Square Error of Dynamic Forecasts
C	0.025
k	0.004
MGS	0.082
XGS	0.081
Y	0.020
P	0.022
$PXGS$	0.031
W	0.033
i_{TIT}	0.240
A	0.022
NFA	0.172
m	0.011
T	0.040
G	0.075
V	0.285
R	0.295
K	0.008
M	0.005
H	0.235
r	0.236
a	0.018
h	0.050

also this variable is highly volatile. Even if the level of the errors is generally higher, the structure is practically unchanged (in the sense that the lowest error variables and the highest error ones are the same in both the in-sample and the out-of-sample case). This can be considered as a favourable feature of the model.

On the whole the results, though not outstanding, look satisfactory[34]. This, in addition to the fact — which we consider more important — that the model has good structural properties (a steady state, stability, sensitivity) and good parameter estimates, suggests an additional examination of its properties through simulation analysis.

---

[34] Comparatively speaking, our results are — at least — not inferior to those obtained by the best among the other continuous time models (RBA 1977, Knight and Wymer 1978) and superior to those obtained by the other continuous model of the Italian economy of a comparable dimension (Tullio 1981).

CHAPTER 4

Policy Simulations

4.1 Introduction

The simulations that we present in the following pages are divided into four groups. The first and most numerous group includes seven antiinflationary policies. With the exception of the first one (the standard monetarist rule) they reflect alternative "ad hoc" antinflationary rules which have been proposed within the Italian economic policy debate and which, in different ways, aim to modify the full wage escalator clauses which were introduced into the Italian economy at the end of 1975.

The second group includes two alternative ways of endogenizing the exchange rate: a freely flexible rate and a monetary authorities' reaction function. The third group assumes a different behaviour for two exogenous variables, the foreign interest rate and world output. The two final simulations consider two alternative global strategies, aimed at curbing inflation and at reducing the public deficit: the first is a shock therapy and the second a gradualist approach.

The simulation exercises we have carried out do not of course exhaust all possibilities. However they display a sufficiently diversified design so as to provide interesting insights into the different properties of the model[1]. For these exercises we used the original non-linear continuous model; details of the procedures are given in App. II, §3.

The results of the simulations are presented as deviations from the control solution. Eight variables are considered: rate of growth of fixed capital $Dlog\, K$, rate of growth of output $Dlog\, Y$, rate of growth of the price of output $Dlog\, P$, the real wage rate $W\!/P$, the balance of

---

[1] Of course this use of simulations is immune from Lucas' (1976) critique. But, in our opinion, this critique is not so cogent as it was thought to be. See Appendix II,§4.

goods and services BGS, defined as the ratio of the value of exports ($PXGS\ XGS$) to the value of imports ($PMGS\ MGS$), the stock of net foreign assets NFA, the stock of international reserves R, the ratio of public consumption in nominal terms to nominal taxes PG/T.

The choice of the variables reflects the emphasis on the medium term properties embodied in our model of an export-led and financially advanced economy, and our idea that, in order to evaluate *ad hoc* policies (as is the case of antiinflationary policies), the overall behaviour of the model, and not just the response of the variables most directly affected by the measures, should be considered.

The first three variables (Dlog$K$, Dlog$P$, Dlog$Y$) are given in tables and figures as differences between the simulation result and the control behaviour; this means that we will talk of an improvement when Dlog$P$ decreases and when Dlog$K$ and Dlog$Y$ increase, and viceversa. The remaining variables are defined as ratios between the simulation result and the control behaviour. This means that we will talk of an improvement when W/P, BGS, R increase and when PG/T and NFA decrease.

Results are presented both graphically and in tabular form. The diagrams are obtained by linking quarterly values while the numbers given in the tables are cumulated yearly values for Dlog$P$, Dlog$K$ and Dlog$Y$ and yearly average percentage changes for the remaining variables. The simulation period runs from 1977-I through 1981-IV.

4.2 Antiinflationary Policies

We have simulated seven alternative antiinflationary policies which, at different times, have been proposed within the framework of the debate on Italian economic policy. We shall consider them separately. It should be stressed that our purpose is not to draw general conclusions as to their comparative efficacy but to consider the results as indicators of the model's behaviour. Each simulation will be discussed both with regard to its effectiveness (i.e. the gain in terms of lower inflation) and its costs which we will identify both in the real wage losses (in which case costs will be borne by wage earners) and in lower growth effects (in which case costs will be borne by the entire community), if and when they occur.

Case 1. The first simulation exercise is the standard monetarist rule. The rate of growth of money supply, m, is kept constant throughout the simulation period at a value which is approximately one percentage point per quarter lower than the control value. We are therefore assuming a considerable monetary squeeze. Contrary to monetarist predictions, the main effect of such a recipe is on quantities rather than on prices. Real output growth is curbed at an increasing rate. The loss in the rate of growth of Y is initially small, but steadly increases (table 12) and reaches a cumulative value of over 0.6 percentage points

Table 12 - Constant rate of growth of money supply

	I	II	III	IV	V
Dlog K	+0.001	-0.062	-0.031	-0.071	-0.141
Dlog Y	-0.034	-0.402	-0.493	-0.544	-0.610
Dlog P	-0.807	+0.643	+0.196	-0.089	+0.031
W/P	+0.089	+0.036	+0.028	+0.091	-0.057
R	+0.097	+1.224	+2.187	+4.418	+7.790
BGS	+0.073	+0.745	+1.806	+2.922	+4.000
PG/T	-0.099	-0.131	-0.098	-0.182	-0.240
NFA	-0.061	-0.763	-0.734	-0.978	-1.472

in the last year of simulation. Lower real growth is reflected in an increasing improvement in BGS which, on average, is 4 percentage points higher in the final year of simulation. Capital outflows are also checked by the lower level of output and hence we obtain an overall increase in the stock of foreign reserves which on average is almost 8 percent higher in the final year of simulation. The fact that the monetary authorities' reaction function is virtually inoperative in this simulation (see appendix II, §II.3) eliminates the positive feedback which a higher level of foreign reserves would exert on the real variables of the economy, both directly and indirectly (via the effect on the stock of bank advances). Let us now turn to the effect on prices.

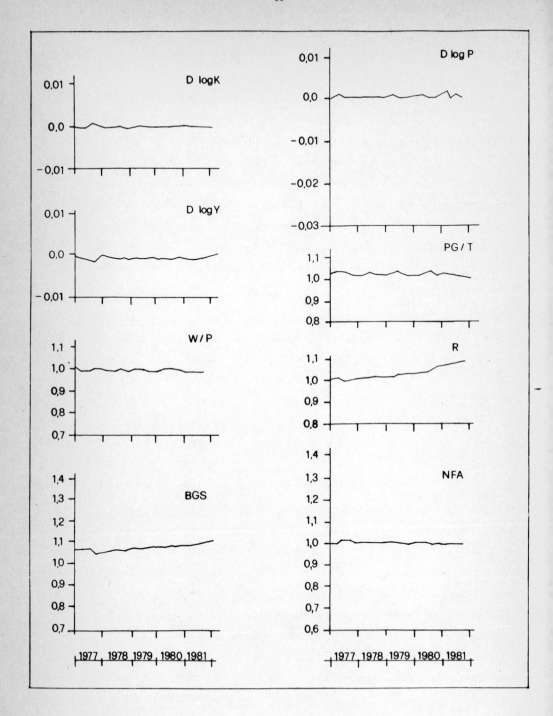

FIG. 1 - Constant rate of growth of money supply

The simulated rate of inflation oscillates above and below its control path; the real wage behaves accordingly, following the behaviour of prices.

The reason for this result will appear more clearly if we look at eqs. (7) and (13.1). If the rate of growth of the money supply is kept constant there is no feedback of the behaviour of (relative) inflation on m (the second term in eq. (13.1)) and hence the effect of m on prices is defined once and for all through the application of the monetarist rule. This means that there is no "buffer effect" of the variations in monetary expansion on prices. This can be explained as follows. The imported element in the formation of the price level $PMGS$, follows an oscillatory trend. This would produce oscillations in the inflation rate if the monetary authorities did not change the value of m (also) according to changes in the inflation rate. As a matter of fact they decrease m when domestic inflation increases and vice versa. Thus *ceteris paribus* this monetary reaction offsets the oscillating impulse which $PMGS$ produces on domestic inflation.

On the other hand the limited magnitude of the oscillations of the rate of inflation, confirms that the relationship between the rates of growth of money and prices is modest as is implicit in the low absolute value of α_{12} and δ_2.

Case 2. The second simulation considers the elimination of "imported inflation" in the wage indexation mechanism. In other words the price index to which money wages are linked — eq. (9.1) — does *not* include the effect of import prices (see appendix for a more detailed description). The gain in inflation is quite strong and increases in the first years. The cumulative gain is more than 4 percent in the second year (table 13), although the inflation rate remains steadily lower than 3 per cent per year with respect to control in the remaining simulation periods. The trade-off in terms of a lower real wage is just as evident. W/P is, on average, almost 19 per cent lower in the first year and keeps on decreasing. In other words, a constantly lower rate of inflation is obtained at the expense of a declining real wage.

Table 13 - Import prices excluded from wage indexation

	I	II	III	IV	V
Dlog K	+0.003	+0.014	+0.061	+0.142	+0.244
Dlog Y	+0.158	+0.567	+0.791	+0.913	+0.999
Dlog P	-3.775	-4.082	-3.245	-3.203	-3.233
W/P	-18.917	-20.502	-21.240	-23.145	-26.611
R	+1.046	+5.105	+7.917	+7.576	+4.657
BGS	+0.387	-0.149	-2.390	-4.871	-7.227
PG/T	-2.088	-4.516	-5.058	-5.229	-5.381
NFA	-0.768	-3.889	-7.187	-9.848	-12.119

The considerable gain in inflation has a favourable impact on capital flows; NFA declines throughout the simulation period. This improvement is largely responsible for the higher level of foreign reserves which increase for the first three years[2]. The current account improves slightly in the first period and subsequently deteriorates increasingly. This deterioration can be ascribed both to quantity and to price effects. The first effect is one of the consequences of the increase in monetary expansion determined by the joint pressure (see eq. (13.1)) of a lower rate of inflation and higher reserves. Higher m (and hence M) means higher consumption and, given the effect of M on bank advances (see eq. (11.1)), higher capital accumulation, which together produce higher output and hence higher imports. Moreover a higher rate of growth in bank advances means lower exports (via α_8). Price effects are contradictory since lower inflation decreases imports but also deteriorates terms of trade. The simulation results would suggest therefore that, in the medium term, negative quantity and terms-of-trade effects prevail over positive competitiveness effects as far as current account behaviour is concerned.

Lower inflation and higher output finally improve the government deficit. PG/T declines steadily and on average is 5 percent lower than

---

[2] The subsequent decline, which nevertheless does not bring the total level of R below its control value, is mainly due to current account behaviour.

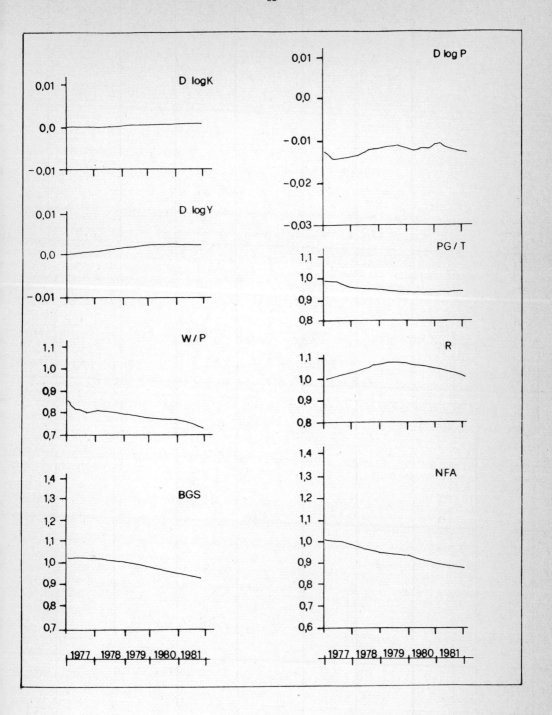

FIG. 2 - Import prices excluded from wage indexation

control in the last three years. Note that this effect does not offset the positive effects of monetary expansion discussed above.

Cases 3 and 4. The following two simulation exercises will be discussed jointly since they differ only slightly from one another. The antiinflationary measure analyzed consists of the predetermination of the rate of increase in money wages throughout the year. This rate of growth varies, i.e. decreases, in the different simulation years. In the first case discussed here, a bonus, which is computed as a percentage increase in money wages equal to the difference between actual inflation and the predetermined rate of growth in wages[3], is handed over to wage earners at the end of each year. If the actual inflation rate is higher than the predetermined rate of growth of wages, the implicit assumption is that firms pay this bonus to workers. In the second case no bonus is handed over to wage earners.

The inflation gain is considerable and increases over time, reaching in the final year a cumulated value of over 3.4 percent (table 14) in the first case and of over 3.6 percent in the second case (table 15).

Table 14 - Predetermined rate of growth of wages with bonus

	I	II	III	IV	V
Dlog K	0.0	+0.001	+0.011	+0.034	+0.074
Dlog Y	+0.020	+0.112	+0.242	+0.389	+0.557
Dlog P	-0.810	-1.480	-2.038	-2.523	-3.413
W/P	-3.291	-6.851	-10.522	-14.235	-19.611
R	+0.125	+0.919	+2.198	+3.449	+4.282
BGS	+0.052	+0.202	+0.313	-1.122	-2.275
PG/T	-0.284	-0.998	-1.899	-2.736	-3.715
NFA	-0.093	-0.711	-1.897	-3.528	-5.542

A considerable qualitative difference in response with respect to the previous simulation exercise emerges (see table 13), where the impact inflation gain was much higher, while the medium-term recovery from

---

[3] See Appendix II, §II.3 for a more detailed description.

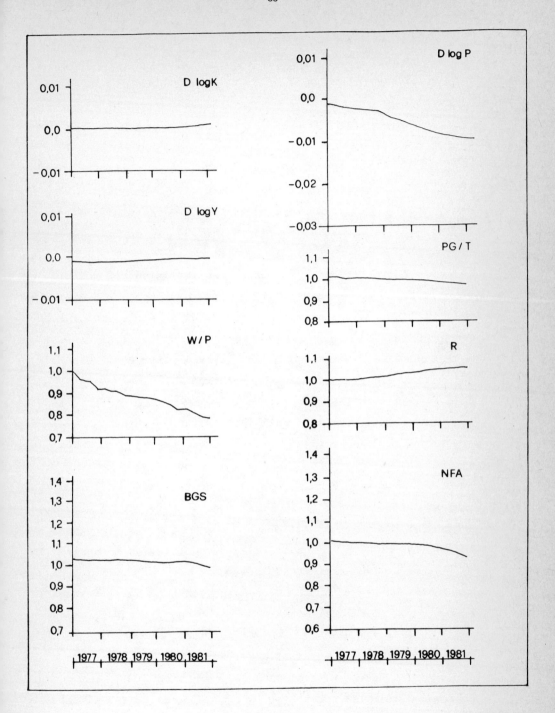

FIG. 3 - Predetermined rate of growth of wages with bonus

Table 15 - Predetermined rate of growth of wages without bonus

	I	II	III	IV	V
Dlog K	0.0	0.0	+0.011	+0.037	+0.076
Dlog Y	+0.020	+0.078	+0.247	+0.401	+0.587
Dlog P	-0.819	-1.507	-2.094	-2.677	-3.622
W/P	-3.327	-6.975	-10.815	-14.890	-20.580
R	+0.125	+0.935	+2.239	+3.549	+4.535
BGS	+0.052	+0.066	-0.310	-1.077	-2.330
PG/T	-0.283	-1.080	-1.930	-2.340	-3.940
NFA	-0.095	-0.725	-1.930	-3.617	-5.760

inflation was practically non-existent as the inflation rate stabilized at a lower level. In the cases under discussion here the impact response is negligible while the medium term path is rather satisfactory. There is no qualitative difference between the two versions considered (with or without bonus) while the quantitative difference is modest as the path profiles closely resemble each other.

The real wage loss follows a path that is symmetrical with respect to that of inflation as it is steadily increasing over time. As one would expect, the real wage loss is higher in the second case (no bonus) but the absolute difference is negligible. It amounts respectively to a little less and to a little more than 20 percent on average in the final simulation year. The fact that the real wage loss is rather consistent in spite of the end of year bonus and of the considerable inflation gain, can be explained if we recall the different motion mechanisms for prices and wages in our model (and indeed in the Italian economy). In continuous time the real wage, W/P, remains constant only if wages and prices grow at the same rate. With the alteration of the wage formation mechanism, eq. (9.1), wages are allowed to rise at a lower rate than the inflation rate, whilst the bonus adjusts the initial value of the money wage but does not take into account the real wage loss which occurs during the year: this is represented by the integral of the difference between the price flow and the wage flow over time. (As P is the output deflator, this is equivalent to the difference of the unitary output

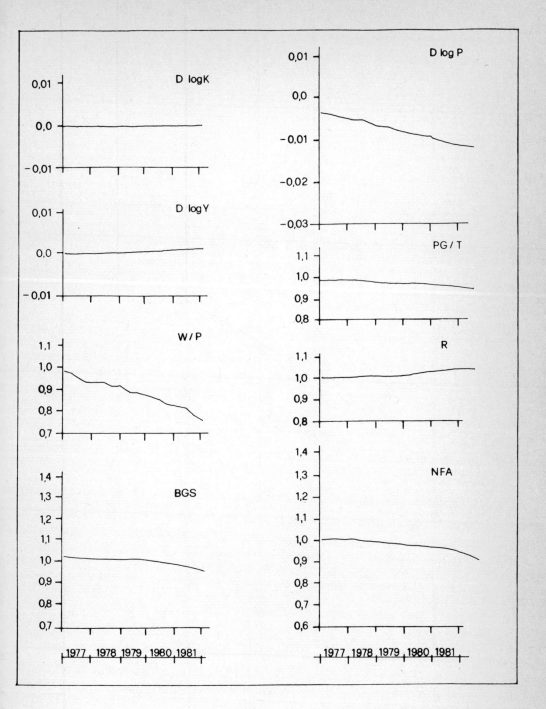

FIG. 4 - Predetermined rate of growth of wages without bonus

flow — for a given rate of growth of real output — and the nominal wage rate).

A word of warning however: this does not necessarily mean that no real wage loss as a consequence of this occurs in the control solution. It simply means that if the wage formation mechanism is altered so as not to take into account this "integral effect", the real wage loss is inevitable. It is worth noting that such an effect can be precisely spotted only if a continuous time analysis is followed. A diagrammatic exposition of this often neglected phenomenon is presented in fig. 5.

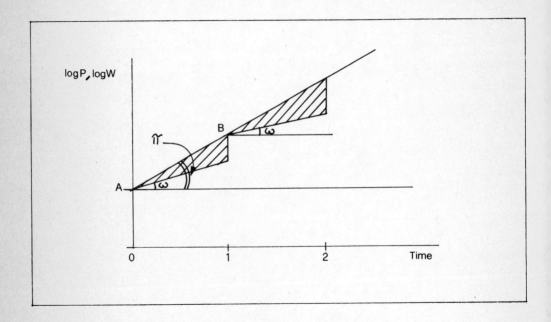

FIG. 5 - Predetermined rate of growth of wages: The "integral" effect

Suppose that at time 0 the real wage is equal to 1 as log W and log P both start at the same point A. Prices rise at a rate equal to tan π while wages rise at a predetermined rate tan $\omega <$ tan π. At time 1 a bonus is handed over to wage earners so that the real wage is reset to its initial level. From $t=1$ to $t=2$ the mechanism is the same as in the previous period and prices and wages rise at the same rate as before. The real wage loss in each period is represented by the shaded area.

The example presented here is an extreme case, since the bonus could be computed so as to have a level of log W higher or lower than log P at point B. The reader may work out intuitively at what level log W should be reset, i.e. the bonus be computed, in order to have a real wage loss equal to zero.

We stress once again that this example is applicable not only to the mechanism we are discussing but to all mechanisms which include a predetermined path of the nominal wage rate with a recovery rule. We have presented the problem at this point since a widespread claim is that such mechanisms do not generally cause real wage losses: a claim which is clearly false[4].

Let us now go back to the effects of these antiinflationary measures on the overall performance of the economy.

The gain in the growth of real output is increasing and becomes particularly evident in the last two years (this gain touches 0.6 percentage points in the no bonus case in the last year). A qualitatively similar gain, but much less pronounced, is present in the rate of growth of K. This modest but not irrelevant inflation/growth trade-off is, as in the previous case, due to the improvement in monetary conditions which in turn are due to the room for expansion following a lower rate of inflation and to the increase in foreign reserves. This last effect in turn is the net outcome of an improving capital account (the stock of net foreign assets decreases steadily) and of a deteriorating current account. The simulated value of BGS is, in both cases, slightly higher than control in the first two years and lower than control in the remaining three.

This sort of "lagged response" in the current account should be as-

---

[4] The same claim, however, is not made by proponents of measures such as those discussed previously: i.e. the elimination of imported inflation from wage formation.

cribed to the lagged gain in real growth (which increases real imports and depresses real exports) and the lagged gain in inflation (which worsens the terms of trade). As in the previous simulation exercise (case 2) the improvement in the real growth/inflation trade-off brings about a similar trend in the value of PG/T which decreases by almost four percent on average in the last year of simulation.

To sum up: the recipe for the predetermined rate of growth of wages performs well in the medium term as benefits are felt on the overall behaviour of the economy at an increasing pace. The higher costs are borne by wage earners who are only partly protected even if a bonus is included (case 3). As a matter of fact the results in the no-bonus case are only slightly more marked with respect to the other case.

Case 5. The next antiinflationary proposal is defined as a "yearly escalator clause", i.e. the mean time-lag of the adjustment of wages to prices is set to four quarters[5] (the estimated mean time-lag is actually around two and a half months). The inflation and real wage response is somewhat similar to that experienced in case 2 (imported inflation excluded from wage indexation). The impact gain in inflation is marked (-2.2 percentage points in the first year, -2.9 percentage points in the second year) but the medium term effect is decreasing although simulated inflation remains well below control inflation (table 16). The

Table 16 - Yearly escalator

	I	II	III	IV	V
Dlog K	+0.001	+0.007	+0.029	+0.075	+0.137
Dlog Y	+0.054	+0.28	+0.48	+0.57	+0.60
Dlog P	-2.21	-2.91	-2.36	-1.85	-1.45
W/P	+8.75	-13.70	-14.50	-14.90	-15.22
R	+0.37	+2.38	+3.71	+5.10	+3.46
BGS	+0.15	+0.09	-1.03	-1.69	-4.33

[5] This does *not* imply that a year is necessary for *full* adjustment of wages to prices, since the mean time-lag is the time needed for about 63 per cent of the difference between actual and desired values to be eliminated.

PG/T	-0.80	+2.50	-3.29	-3.37	-3.17
NFA	-0.27	-1.83	-4.05	-6.00	-7.36

real wage follows a pattern already witnessed in case 2. The loss in W/P is substantial right from the first year (almost 9 percent) and then slowly it increases; it finally stabilizes around 15 percent in the final period. The loss in real wage occurring in this case is again due to the "integral effect" discussed above, although the mechanism which produces this effect is different. The growth rate of wages is not predetermined and hence the target value of wages is the same as in the control solution (contrary to what was implicit in the previous simulations where the value of $\widehat{W}$ was changed with respect to control); however this unchanged target level of W is reached over a much longer time than in the control solution and this very fact determines the loss in the real wage over the same period. However, once the new adjustment mechanism has had time to establish itself, the real wage loss with respect to control does not vary considerably.

Lower inflation again produces higher output growth. The gain is especially marked after the second year of simulation and it reaches a 0.6 percentage improvement in the final year when real capital growth shows a slight increase with respect to control.

The transmission mechanism of lower inflation to higher growth is, once again, monetary policy. Lower inflation and higher international reserves allow for higher monetary and hence credit expansion. This expansionary effect however loses some of its strength in the final simulation period (and this is reflected in the decrease in the real growth gain mentioned above). This is due to the decreasing inflationary gain in the final period of simulation (which induces a more restrictive monetary policy) and to the decreasing gain in foreign reserves. This second effect in turn is the net result of two contrasting forces: 1) a gain in capital outflow performance (as in previous simulations), which, however, slows down (increases less rapidly) after the third year of simulation (and this again is a consequence of a lower inflation gain *and* of higher real growth); 2) a deterioration in the

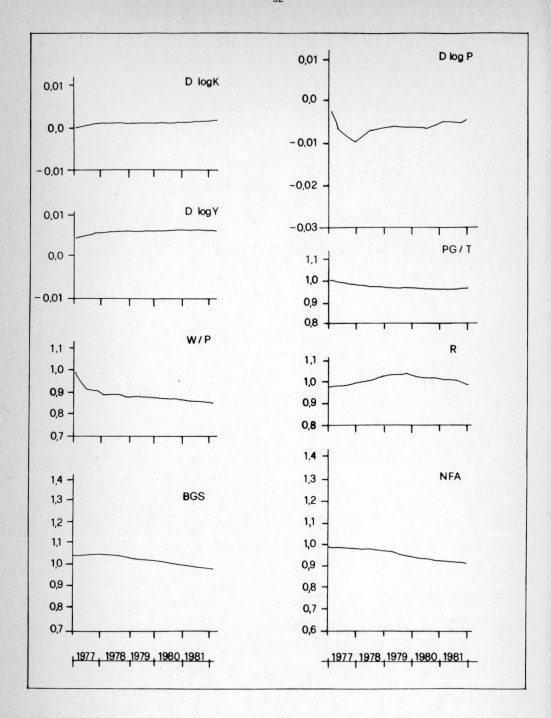

FIG. 6 - Yearly escalator

current account which is rather marked in the final period of simulation.

The gain in government deficit performance reaches a peak after the third year of simulation and then stabilizes on yearly averages.

Case 6. The principle underlying antiinflationary recipes discussed in cases 2, 3 and 4 was the change in the value of the target money wage, $\widehat{W}$, while in case 5 the target wage was untouched and emphasis was placed on the adjustment mechanism. This simulation, on the contrary, assumes a given real wage as the leading principle. The recipe assumes a given rate of change in the money wage every year (this rate is assumed to be zero in our case). At the end of the year wage earners are given a bonus which includes two components. A first component brings the nominal wage to a level which, given the new price level, sets the real wage at the level it had at the beginning of the simulation period; a second component should take into account the real wage loss due to the integral effect discussed above. Fig. 7 depicts a diagrammatic presentation of the proposal. At $t=1$, $\log W$ is moved from C to B (and this sets the real wage back to its initial value which is here supposed to be one).

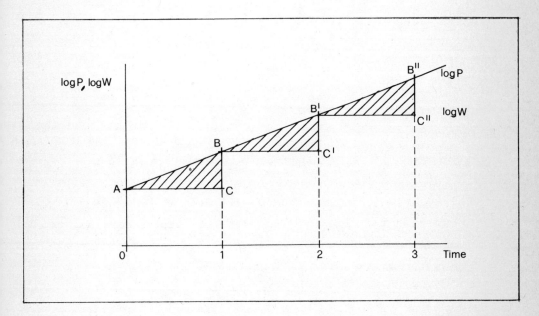

FIG. 7 - Constant money wage and yearly escalator with bonus: A description

In addition to this, wage earners are given a bonus which is equal to the shaded area ABC. The original proposal assumes that this additional bonus can be handed out to wage earners gradually so that it really amounts to an increase in the nominal wage rate and that interest should be paid to wage earners to compensate for the loss suffered during that period. The way in which this proposal was formalized is described in detail in appendix II, §II.3. The degree of approximation explains why this simulation does not produce exactly the predicted results. As a matter of fact, the real wage decreases on a yearly average from minus 6.03 percent in the first year of simulation to minus 16.03 percent in the final year (see table 17). In this case however, it is

Table 17 - Constant money wage and yearly escalator with bonus

	I	II	III	IV	V
Dlog K	+0.001	+0.003	+0.017	+0.041	+0.078
Dlog Y	+0.044	+0.151	+0.25	+0.34	+0.43
Dlog P	-1.49	-1.77	-1.80	-2.00	-2.70
W/P	-6.03	-8.32	-9.88	-12.10	-16.03
R	+0.28	+1.32	+2.37	+2.88	+2.70
BGS	+0.11	-0.02	-0.59	-1.40	-2.37
PG/T	-0.57	-1.31	-1.80	-2.14	-2.70
NFA	-0.21	-1.04	-2.19	-3.44	-4.80

interesting to look closely at the time paths of W/P too. As can be seen in fig. 8 , the real wage increases at the beginning of each year of simulation and then falls following a downward trend. As a matter of fact the cause for this less than full real wage protection is the approximation with which the second component of the bonus (i.e. the one which should cover the loss due to the integral effect) was computed. Nevertheless the simulation exercise well represents the overall effects of the proposal. The above mentioned effect of the real wage can be better understood if one looks at the inflation gain. This is on average increasing and reaches 2.7 points in the last year of simulation; the time path behaviour is again different. The inflation gain

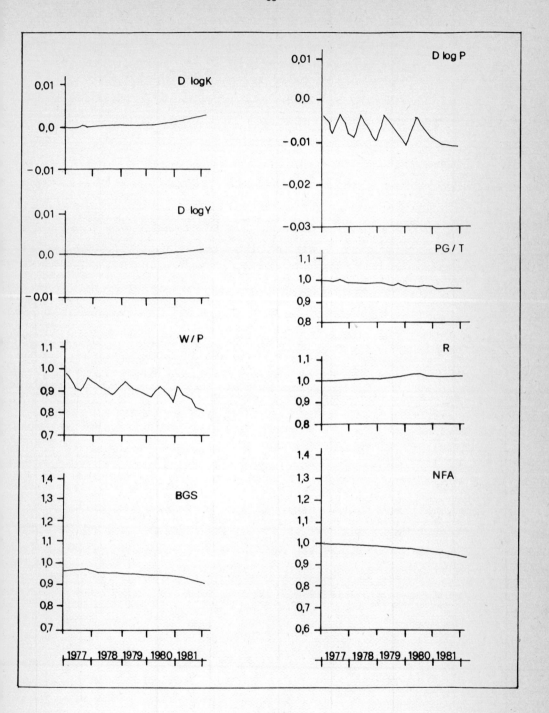

FIG. 8 - Constant money wage and yearly escalator with bonus

is at a maximum at the end of each year of simulation and then practically disappears at the beginning of each new year i.e. when the money wage is readjusted to take account of the real wage target discussed above. The money wage/price link in other words is, in this simulation, particularly close on account of the real wage target. It is worth noting that this rather strong oscillatory motion in prices and wages is not transmitted to the other variables considered here. The reason is that the monetary authorities' reaction function provides the buffer by altering the growth rate of the money supply. The effects on the overall behaviour of the economy are somewhat less pronounced in this case. Real growth gains are present (both in output and in capital accumulation) but are smaller than in other simulations. The transmission mechanism is, once again, the monetary expansion which lower inflation allows. The usual foreign reserves build-up is present and increases up to the fourth year of simulation after which it declines slightly. This is the result of a marked capital inflow (decrease in *NFA*) and a relatively high deterioration in the current account. Lower monetary growth should also be ascribed to the gain in government deficit.

Case 7. We now come to the final simulation exercise included in the first group: the price and wage freeze. As is well known and as many historical experiences have confirmed[6], the main problem in the implementation of such a policy is the way out of it.

We tried to deal with this problem by designing a three-stage policy. The first stage covers the first two quarters during which the growth rate of the price of output and of money wages is set to zero. Export prices, on the contrary, are left as in the original version. During the following two quarters (stage two) the price of output is allowed to rise only by the amount due to the rise in import prices, which, in eq. (7.1), are a proxy of the price of imported inputs, while nominal wages are still kept constant. The rationale for this is that the way out of a price and wage freeze should take into account mainly the in-

---

[6] For a theoretical treatment of the problem and for historical records see Galbraith (1952) and Meade (1957).

come distribution effects that such a policy implies. In our case we assume that in phase one profits are curtailed (since, apart from money wages, the costs of imported inputs may continue to rise) while in phase two wage earners bear the costs since real wages are reduced.
During the four years of phase three, the model is set back to its original structure.

Table 18 - Price and wage freeze

	I	II	III	IV	V
Dlog K	+0.009	+0.032	+0.11	+0.16	+0.18
Dlog Y	+0.48	+0.75	+0.32	+0.06	-0.10
Dlog P	-9.90	+3.10	+2.25	+1.54	+1.06
W/P	-2.47	+1.70	+1.60	+1.09	+0.75
R	+3.36	+8.34	+3.68	-3.09	-8.24
BGS	+0.97	-1.99	-4.71	-4.85	-4.06
PG/T	-5.46	-3.68	-0.18	+1.10	+1.35
NFA	-2.41	-6.61	-6.11	-4.23	-2.55

Simulation results confirm our expectations. The inflationary gain is dramatic in the first year (almost 10 percent) but is already negative (i.e. higher rate of inflation) in the second year as inflation rises three percentage points above control. Inflation is afterwards constantly above control although this inflationary loss is decreasing and settles at one percentage point in the final year. From this point of view, then, the implementation of such a policy is absolutely unsatisfactory. Let us now look at the effects on W/P. The real wage loss is substantial but not dramatic (2.5 percentage points) in the first year. What might appear surprising at first sight is that the real wage is on average higher than control for the rest of the period. This is explained by the fact that, apart from the first year, the mechanism of wage formation is untouched. This means that during the first two phases the discrepancy between the actual and the desired level of the money wage rises considerably, so that when the mechanism is again free to operate, the growth rate of money wages exceeds the growth rate of

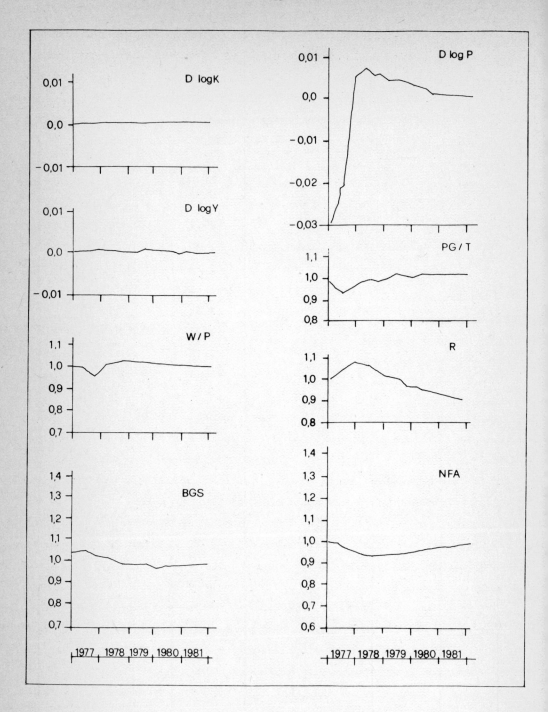

FIG. 9 - Price and wage freeze

prices in the quarters immediately following the end of phase two and this produces an overshooting of wages with respect to the control equilibrium value. This, in turn, accelerates the wage-price spiral and hence produces the inflationary loss we have mentioned.

The inflationary gain produces positive effects on international reserves. R rises in the first two years and then declines becoming lower than control in the two final years. This lagged effect of inflation on foreign reserves is the outcome of a negative effect on the current account. The considerable inflation gain in the first year produces a small improvement in BGS in that period (around 1 percent higher than control) but we then have a deterioration which is maintained at a rather constant level in the last three years. This deterioration can be imputed mainly to the loss of competitiveness and hence higher imports following a constantly higher rate of inflation when the time lags have been allowed for. Capital outflow benefits from the initially lower inflation; the level of NFA decreases with respect to control for the first three years and then starts to increase again in the two final years although it keeps below control.

This medium term drain in reserves, together with the medium term inflationary loss produces a monetary squeeze which is largely responsible for the medium term deflationary effect on real output growth. This is lower than control, albeit slightly, in the final simulation year after having been higher but with a decreasing trend in the previous years. The effect on government deficit is encouraging only if one looks at the impact effect. The gain in PG/T is marked only in the first two years; it is then decreasing and eventually becomes a loss (higher PG/T) in the last two years. Finally we may note that this simulation exercise has confirmed the importance of the major transmission mechanisms i.e. the crucial role of monetary policy in linking price and quantity variables which operate in the model. In fact, what has turned out to be a higher inflation scenario has made these mechanisms operate in the opposite way with respect to the previous exercises.

We may conclude this section with a few general considerations on the results reached so far. As we said at the beginning, it is not our intention to pick out the "best" antiinflationary recipe from those tested (the reader may chose his favourite one if he wishes). As a matter of fact on the basis of the results discussed above, we may state that there is no single recipe which stands out as the best one if one takes into consideration the overall costs and benefits (whereas we would not hesitate in awarding the title of "first-worst" to the price and wage freeze proposal). These exercises however have shed some light on the overall behaviour of our model. First of all, a strong real wage-inflation trade-off has emerged and this is not surprising if one recalls the way in which inflation is modelled [see eq. (7)] and that most of the policies tested assume that in order to curb inflation the wage formation mechanism must be altered in some way or another. (Other results on the antiinflationary front may be reached if, as is the case in the next two simulations, the imported component of inflation is lowered through an exchange rate appreciation).

Secondly, the behaviour of inflation does have an impact on the real variables if one takes into account the full real-monetary links present in the model. Lower inflation has a beneficial effect on capital outflows which are positively related to price increases. The effect on the current account is however ambiguous since the gain in competitiveness is offset by terms-of-trade deterioration; the overall effect may be enhanced in one way or another according to the real growth effect.

In general our results show that lower inflation has a beneficial effect on foreign reserves and this result is, in itself, a sufficient incentive for undertaking antiinflationary policies. Contrary to widespread views, this beneficial effect derives mainly from capital account performance, an aspect which is often neglected in policy debates.

Real growth performance is strongly affected by the consequences which inflation produces on monetary (and hence credit) expansion. In addition to the increase in foreign reserves a lower rate of inflation is associated with a higher rate of growth in the money supply since the monetary authorities react positively (i.e. they expand the money supply) whenever inflation decreases.

Lastly, higher real growth and a lower rate of inflation improve (decrease) the government deficit and this constitutes another incentive for pursuing such a policy. No clear evidence however can be found of the inverse relationship. In fact, a smaller deficit can help to curb inflation only through its effect on money supply which may well be more than offset by the behaviour of other variables. This aspect will be discussed again when overall strategies (cases 12 and 13) are examined.

4.3 Endogenous Exchange Rate Determination

The next two simulations will be discussed together as they differ only slightly. In both cases the exchange rate is endogenously determined. This implies that a) foreign variables which are expressed in lire will be expressed as the product of the (lira/US dollar) exchange rate and the value of the variable expressed in dollars, b) a new equation is added to the model for simulation purposes.

Case 8. The first simulation is a managed float exercise, where we assume that the monetary authorities set a target exchange rate as a function of export (price) competitiveness. Thus the exchange rate is devalued whenever the rate of growth of export prices exceeds the rate of growth of world prices, PF, account being taken of the change in foreign reserves, i.e. we assume that the monetary authorities have a target value for the stock of reserves. The adjustment speed is set to a value which is equal to α_{20} i.e. the value of the monetary reaction function adjustment speed.

Since the simulation period (1977 to 1981) is — historically — one of managed float with respect to the dollar, the observed value of the exchange rate which is included in the data, incorporates the effect of the intervention policy of the Bank of Italy. This exercise therefore amounts to assuming a different intervention rule with respect to the one which was effectively followed. As we mentioned in the previous chapter the switch in the exchange rate regime would have required the use of nonlinear methods for the estimation of the monetary authorities' reaction function. Estimation results have proved to be acceptable all the same and this may be interpreted as the fact that the behavioural assumptions embodied in the parameters remained rather stable throughout the sample period. This relative "robustness" allows us to hypothesize at least as an approximation that the monetary authorities' reaction function may be left unchanged also in the present simulation exercise, i.e. that although the intervention rule has changed the monetary policy rule has not. Let us now discuss the simulation results. Although exchange rate behaviour is not represented in our tables a faithful idea can be drawn from the behaviour of inflation (table 19).

Table 19 - Managed float

	I	II	III	IV	V
Dlog K	-0.04	-0.12	-0.42	-0.68	-0.81
Dlog Y	-1.99	-2.89	-2.07	+0.30	-0.95
Dlog P	-5.75	-8.91	-8.20	-4.13	-6.00
W/P	+1.50	+4.50	+7.47	+8.94	+11.00
R	+2.03	-26.70	-56.10	-69.00	-77.00
BGS	+15.00	-1.23	-11.20	-10.20	-7.15
PG/T	-2.35	-7.50	-11.30	-11.45	-11.50
NFA	+10.10	+32.00	+45.00	+37.80	+38.10

During the first two years of simulation a considerable inflationary gain occurs; this decreases in the third year, increases in the fourth year and then rapidly vanishes in the final quarters. This inflationary behaviour is (partly) associated with exchange rate movements (in the sense that the latter determine the former and not viceversa). Decreasing inflation then means an appreciating currency and increasing inflation means a depreciating currency. This amounts to say that this kind of intervention rule would produce a strong tendency to revaluation at least in the first phase of the experiment. The explanation for this outcome is to be found in the deflationary effect that an unchanged level of foreign reserves produces on the economy via its negative effect on money supply. This we may assume as the impact effect; from

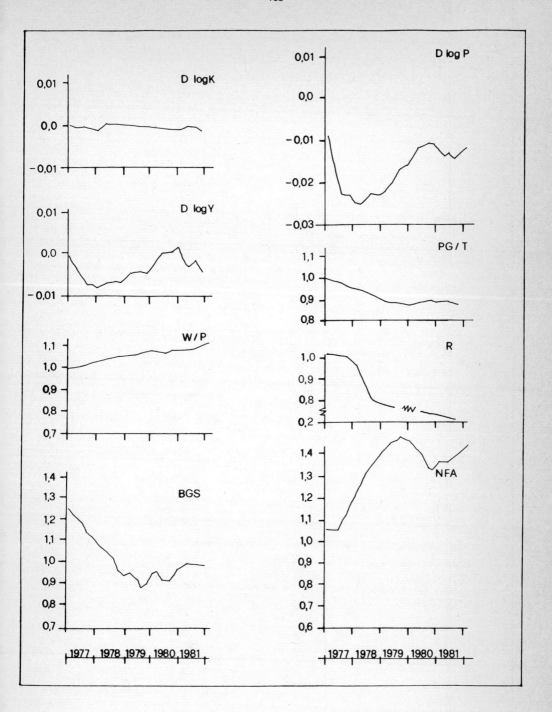

FIG. 10 - Managed float

this moment onwards, the response of the model is rather complicated as it involves a full interaction of monetary and real variables. For expository purposes we shall divide the description of the mechanism into three distinct phases. In phase one the deflationary effect of lower money growth produces a dramatic improvement in the current account as the quantities of exports are increased, both as a consequence of the credit squeeze which takes place and as a consequence of the lower growth of output, which in turns is generated by the lower level of consumption produced by the monetary squeeze.

The fall in real growth also heavily curbs the quantities of imports. Monetary deflation, in addition, helps to bring down inflation. The dramatic improvement in the current account induces a revaluation in the exchange rate in order to check the increase in reserves. This brings us to phase two which is characterized by a deterioration of the current account which is due; a) to increased imports which follow terms of trade improvement determined by the revaluation, b) to a pick up in the growth of output which is a consequence of exceptional export growth in the previous phase (note that, in this exercise, the J curve effect works in the opposite direction, as we are witnessing a revaluation which initially improves but subsequently deteriorates the current account). Both the strong revaluation and the pick up in real growth are responsible for the increase in net capital outflows which takes place during these two phases. During this second phase the general tendency of DR is to decrease, hence the exchange rate responds with a devaluation which, as we mentioned above, is reflected in higher inflation. This brings us to phase three. During this last phase the current account again improves thanks to increased competitiveness and to lower real growth which comes as a result of lower exports. This latter outcome is (only slightly though) the effect of lower prices competitiveness deriving from the previous period's revaluation and, mostly, the effect of negative supply constraints, which are the joint result of higher output growth and of a negative rate of growth of capital stock due to the credit squeeze. A negative value of $D\log K$ is a constant feature in this simulation, but its effects on exports are felt with a considerable lag since the mean time-lag of the stock of capital is five years. Lower growth and devaluation of the exchange rate produce a positive effect on NFA which tends to decrease. This phase is then characterized by a tendency of foreign reserves to increase; the economy is therefore ready for a new phase (which we will not discuss here) in which the exchange rate will tend to appreciate. The behaviour of PG/T appears much smoother than that of other variables. This should not, however, be considered as surprising since this result (which implies a rather steady lower deficit during the simulation period) is the outcome of two opposing tendencies. As previous simulations have shown, higher real growth and lower inflation improve (i.e. lower) the deficit and viceversa. Now, it can be seen that, in each of the phases, these two forces have worked so as to offset each other: lower growth has been accompanied by lower inflation and viceversa and this explains the apparently odd (i.e. stable) behaviour of PG/T.

This simulation allows us also to appreciate the effects of a revaluation on inflation. As we have already seen inflation is below control for the whole period although its behaviour is somewhat cyclical. This should be ascribed to the operation of two factors: the monetary squeeze, which keeps inflation low, and the exchange rate movement, which is responsable for the cyclical behaviour. This also means that, in this exercise, the causal relationship between P and W is somehow inverted. Lower inflation produces lower money wages and not viceversa (since, unlike what happened in previous exercises, the equation which determines W is not modified by any policy intervention). This, in turn, produces a higher real wage because of lower inflation. A result which is opposite to that achieved in previous exercises. To put it again differently, in the previous simulations the main cost of lower inflation was a lower real wage (and a trade-off inflation-growth was found), in this exercise the main cost is lower growth (and a trade-off inflation-real wage is found).

We will conclude the discussion of this simulation by stressing what follows: the model's response to a change in the exchange rate regime may be split into two effects. An impact effect, which is the response to a much lower growth in foreign reserves; this effect is highly de-

flationary and operates mainly through quantity reactions. A lagged effect, which is the response to the change in the exchange rate and hence in prices. These two effects enhance each other and, as we have seen, are present in the different phases of the simulation discussed above.
Case 9. The rather lengthy description of the managed float case allows us to go rapidly through the flexible rate case. In this exercise the exchange rate varies in response to the excess demand for foreign currency (measured by the overall balance of payments). Since no official intervention is assumed the exchange rate equation is specified (see appendix II, §II.3, for a full description) so as to have the actual exchange rate adjust to its partial equilibrium level with a mean time lag of a day (the adjustment speed is set equal to 90). This level is defined as that rate which clears the foreign currency market (hence the exchange rate at which $DR=0$). It should be stressed that, by so doing, we do not necessarily accept the "popular" or traditional flow hypothesis of exchange rate determination. In fact, what we do is to mimic the functioning of a freely flexible exchange rate regime, where exchange rates move to equilibrate supplies of and demands for currencies. As a general statement, this view is uniformly accepted by economists today (Isard, 1978, p. 8; Kouri, 1983, p. 116), and does not imply adhesion to any particular view as to the markets behind these supplies and demands.

In this respect this simulation exercise has proved to be a success. The level of foreign reserves which results at the end of the simulation period is just 0.4% higher than the value at the beginning of the period; this appears as a steady (and substantial) drop in the level of R with respect to the control value, which shows an increasing trend. The constant value of R is associated with a rather pronounced exchange rate fluctuation, as one would expect.

The behaviour of the model in the freely flexible case is substantially similar to that of the previous case. The variables respond, however, more sharply and, of course, we do not witness the (slight) improvement in foreign reserves. This also explains the somewhat greater deflationary impact with respect to the previous case.

It is generally assumed that, given perfect flexibility, monetary

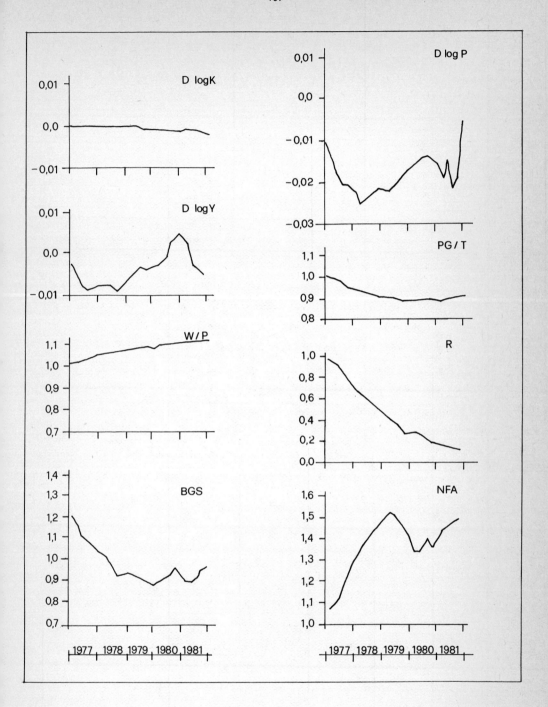

FIG. 11 - Freely flexible exchange rate

Table 20 - Freely flexible exchange rate

	I	II	III	IV	V
Dlog K	-0.05	-0.18	-0.47	-0.72	-0.82
Dlog Y	-2.45	-2.80	-1.98	+0.93	-1.35
Dlog P	-6.61	-9.03	-8.03	-4.05	-5.71
W/P	+1.90	+5.00	+7.95	+9.25	+11.50
R	-7.59	-35.00	-62.00	-72.00	-81.00
BGS	+12.00	-2.50	-12.60	-10.70	-10.00
PG/T	-3.10	-8.10	-11.70	-11.50	-11.80
NFA	+13.50	+35.00	+46.90	+36.50	+40.40

policy gains additional freedom, and hence one could suggest that the monetary authorities reaction function should be modified accordingly. We decided, however, to leave the reaction function unchanged, i.e. to assume the same monetary policy rules, for two reasons. In the first place, as we have said, the simulation results for the freely flexible rate case are not qualitatively different from those obtained in the case of the managed float; as a consequence, the same reasons for not altering the monetary function apply in this case. Secondly, although free flexibility may imply greater freedom of action, it does not necessarily imply a change in the targets of the central bank as they are embodied in eq. (13.1).

4.4 Exogenous Variables Behaviour Altered

The next two simulations consider the reaction of the model to a different behaviour of two exogenous variables: the foreign interest rate i_f and foreign demand proxied by YF. Both simulations assume a more expansionary and more stable international environment; in the first case the foreign interest rate is constant for the whole simulation period at the level it assumed in the first year of simulation.

Case 10. The limited response of the model reflects the estimated re-

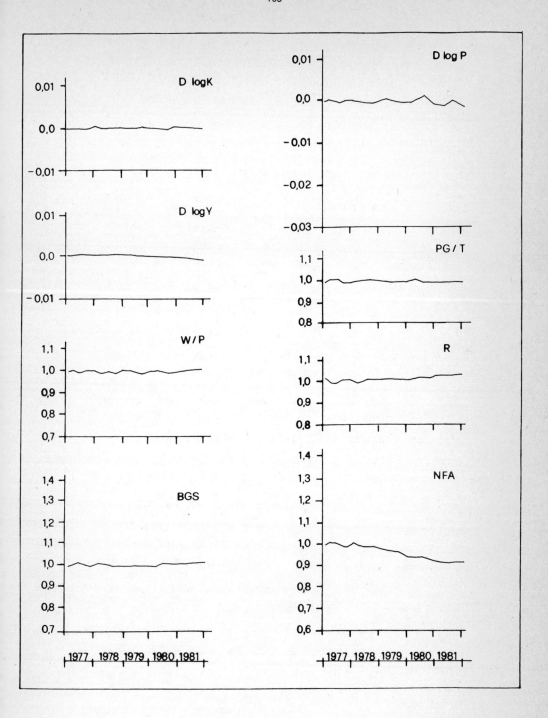

FIG. 12 - Constant foreign interest rate

sults. As it will be recalled (see ch. 2), the foreign interest rate influences the rest of the model through its effects on net foreign assets and through the domestic interest rate. This variable, in turn, affects significantly (and negatively) only the expansion of bank advances. The only noteworthy result of this exercise is a lower stock of net foreign assets. This effect is, however, partially offset by

Table 21 - Constant foreign interest rate

	I	II	III	IV	V
Dlog K	0.0	0.0	+0.015	+0.011	+0.004
Dlog Y	0.0	+0.001	-0.010	-0.07	-0.11
Dlog P	0.0	+0.004	-0.002	0.0	-0.015
W/P	0.0	-0.1	-0.1	-0.1	+0.1
R	-0.1	+0.2	+2.1	+6.0	+8.6
BGS	0.0	-0.1	-0.2	-0.1	+0.3
PG/T	-0.1	0.0	-0.1	-0.1	-0.1
NFA	0.0	-0.1	-0.5	-5.1	-8.0

a higher credit expansion so that only in the last two years the level of NFA decreases appreciably. This is reflected in an accumulation of foreign reserves which is relevant only in the final part of the simulation period. The relatively pronounced effect of credit expansion is reflected in the rate of growth of capital stock which is slightly higher than real output growth (although the absolute values are negligible). This result is a novelty with respect to the simulation exercises previously considered (the gain in Dlog$Y$ is usually higher than the gain in Dlog$K$ when an expansionary effect prevails) and it stresses the fact that the simulation produces a credit expansion which is not determined by a money supply expansion; but rather it is the credit expansion which determines the improvement in foreign reserves. It shows, in other words, the autonomous response of the banking sector — eq. (11.1) — .

Case 11. The reaction to a higher and more stable world demand (the rate of growth of YF is set at 9 percent per year throughout the period) is produced mainly by the behaviour of real variables.

As one would expect the simulated path of YF enhances the export-led characteristics of the model. This is reflected in a considerable gain in output growth and in an improvement in the current account. This also means that an export impulse does not produce a high enough increase in imports to do more than just offset the gain in exports. This result is due only to quantity effects, as throughout the simulation period the net gain (or loss) in inflation with respect to control and hence in terms-of-trade is negligible; as a consequence, the real wage is practically unaffected. The examination of this export-led growth exercise brings to light some interesting medium-term properties of the model. The impact effect of higher exports is quite strong in the first two years; the following two years however display a lower gain in the rate growth of output which decreases by almost one percentage point. Dlog$Y$ however picks up in the final year of simulation passing again the 4 percent per year value. The explanation for this behaviour lies

Table 22 - Higher world demand

	I	II	III	IV	V
Dlog K	+0.04	+0.31	+0.78	+1.21	+1.57
Dlog Y	+4.18	+4.20	+3.29	+3.07	+4.07
Dlog P	-0.17	+0.02	+0.01	0.00	-0.013
W/P	+0.01	+0.003	0.00	-0.01	0.00
R	+44.3	+105.90	+142.00	+170.10	+237.80
BGS	+23.70	+24.10	+20.70	+22.70	+32.70
PG/T	+0.17	+1.08	+2.03	+2.01	+3.14
NFA	-20.70	-34.80	-39.20	-42.10	-48.50

in the export, output, capital stock (supply constraint) link embodied in the model. Initial high export growth stimulates output growth and this produces a *negative* effect through the supply constraint on exports -- see eq. (5.1). Output growth however increases the growth

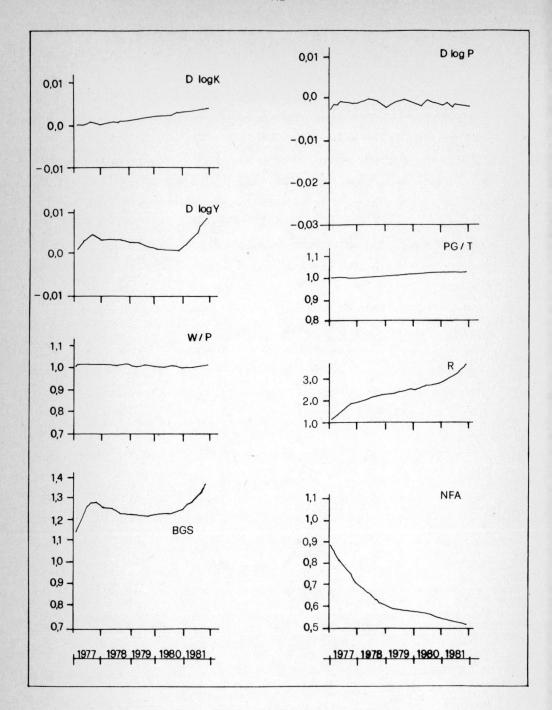

FIG. 13 - Higher world demand

in fixed capital stock whose reaction has a mean time-lag of five years. This means that entrepreneurs will take such a lag to change their desired value of K. Therefore, only after such a lag will productive capacity be adequate to circumstances. This explains why, given a strong world demand, exports (and hence output) may again rise after the downturn.

Output growth and real accumulation growth are also enhanced by the considerable monetary and credit expansion which follows the rise in foreign reserves. This in turn is the result of an improvement in both the current and the capital account. NFA display a lower and steadily decreasing level throughout he period due to the higher level of YF — see eq. (12.1) — which offsets both the higher level of Y and the larger credit expansion.

4.5 Global Strategies

The two final simulation exercises consider two global policy strategies. Their overall dimension stems from the fact that all policy channels present in the model are simultaneously activated: their difference lies in the tuning chosen for the implementation of the strategies. The first one is our interpretation of a shock therapy treatment while the second is a gradualist approach. The goal of both policies is to try to drive an inflation and government deficit prone economy, like the Italian one, back on to an equilibrium path[7].

Case 12. The shock therapy strategy includes the following policy measures: a) a once-and-for-all increase in taxes and a once-and-for-all decrease in government expenditure so that the resulting government deficit (PG/T) is brought back to the European Community average level; b) price and wage freeze (i.e. the policy discussed under case 7) after which a yearly escalator is introduced as in case 5; c) a once-and-for-

---

[7] The shock therapy vs. gradualism debate has recently been brought to general attention. See for a discussion and different points of views, Fellner *et al.* (1981).

all increase in the stock of credit (+20.0 percent); d) capital outflows are frozen (almost) completely for one year; e) the exchange rate is endogenized (an exchange rate equation is added to the model as in cases 8 and 9. A once-and-for-all devaluation of 30 percent is followed by the imposition of a fixed rate) during the whole period[8]. We will now discuss the effects of such a strategy on the model by considering the impact effects first. The public finance pressures sub a) produce a once-and-for-all improvement in the government deficit; PG/T improves by over 20 percent and this improvement is, more or less, maintained for the whole simulation period. The 30 percent devaluation produces

Table 23 - Shock therapy

	I	II	III	IV	V
Dlog K	-0.025	+0.103	+0.26	+0.25	+0.18
Dlog Y	+0.46	+0.38	+0.023	-0.33	-1.74
Dlog P	-6.71	+7.75	+4.10	+2.63	-1.50
W/P	-3.65	-11.25	-13.15	-13.95	-13.20
R	+600.00	+472.50	+65.00	+95.00	+215.00
BGS	+38.50	+56.90	+56.90	+62.30	+87.50
PG/T	-23.70	-21.30	-18.90	-18.60	-20.20
NFA	-91.03	-87.40	-59.90	-39.00	-25.00

initial improvement in the current account. This dramatic improvement however is largely due to the decrease in the *level* of output which results from the once-and-for-all improvement in public finances (as a matter of fact the level of output with respect to control decreases by 0.7%). The price and wage freeze produces an initial inflationary gain of 6.7 percent in the first year of simulation.

This price freeze initially protects the economy from the initially

---

[8] Note that a fixed rate does *not* mean that the exchange rate assumes its original (control) value which in turn follows historical values as the exchange rate is exogenous in the control solution. Instead, we assume that the rate of change of E is 0.0 throughout the period, which means a revaluation with respect to control.

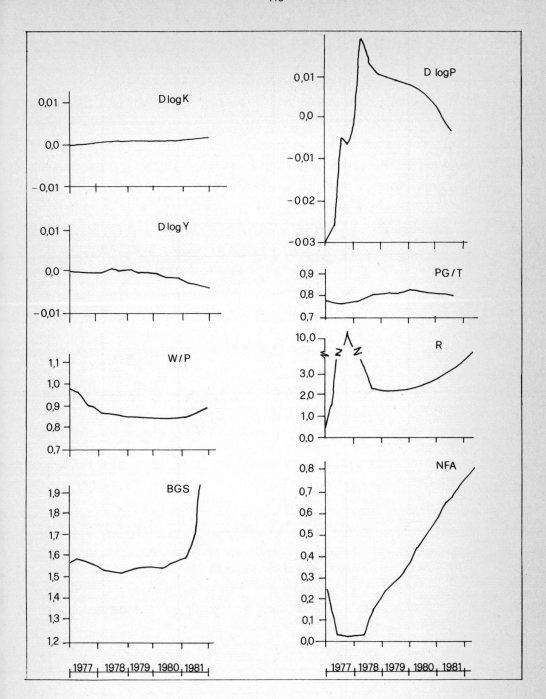

FIG. 14 - Shock therapy

in imported inflation due to the one shot devaluation. The real wage falls in the first year by almost 3.7 percent. The freeze in capital outflows produces a dramatic improvement in *NFA* and this result, together with the improvement in the current account, determines an extraordinarily large increase in foreign reserves. Note that the choice to sterilize almost totally capital outflows is made necessary, among other reasons, to offset the positive effect that the once-and-for-all increase in the stock of bank advances would have produced on *NFA*. This, in turn, is a necessary measure if we wish to avoid debt deflation effects associated with shock policies (see Fischer, 1981). Let us now discuss the behaviour of the *growth rate* of output and of fixed capital stock and, with them, the medium term response of the model. The growth rate of output increases slightly over control but in spite of this upward trend it falls below control in the two final years. This behaviour may be explained as follows. The gain in Dlog$Y$ in the first part of the simulation should be ascribed to the combination of three elements.   i) The rise in real exports favoured (only partly) by the initial devaluation and largely by the above mentioned fall in the *level* of output which eases supply constraints on exports. ii) The one shot increase in the stock of credit, on the other hand, depresses exports but increases capital accumulation and then increases output and subsequently exports as it further eases the supply constraint on *XGS* while the depressive effect of the increase in A fades aways. iii) The third initial expansionary effect comes from the growth in money supply which results from the initial inflationary gain and from the domestic increase in foreign reserves.

After about three years, however, the expansionary forces give way to depressive ones. The fixed exchange rate which weakens competitiveness,and previous output and monetary growth (via further credit increases) slow down the initial export drive. But the most serious constraint to growth comes from an inversion in general monetary conditions. A slow down in money supply growth originates in three different channels. The first one is the gradual decrease in the accumulation of foreign reserves due to the large increase in capital outflows which gradually return to the value they assume in the control solution. The second one is the increase in inflation which not only accounts for the "spring back" effect which follows — as we have seen in case 7 — the way out of price freeze measures, but also includes the higher imported inflation due to the initial devaluation. The third effect comes from the stabilization in the improvement in the government deficit which gradually more than offsets the increase in monetary creation due to foreign reserves growth (remember also that δ_3, the weight associated with the rate of change in the public sector's borrowing requirement in the monetary authorities' reaction function, is larger than δ_4, the weight associated with the rate of change in foreign reserves). The monetary restriction which results, negatively affects real consumption and hence output. This deflationary tendency is also the main cause of the improvement in the current account which takes place in the last two years of simulation in spite of the realtive decline in export growth.

A final word on inflation and wages. The effects mentioned above produce a gain in inflation in the final year only as a result of three elements: the progressive monetary squeeze, the slow down in the increase in foreign inflation thanks to the fixed exchange rate and the effect of the yearly escalator on the behaviour of nominal wages. Real wages show a loss with respect to control which increases in the first two years and then stabilizes for the remaining period. This is the result we obtained in the yearly escalator exercise (case 5). The relative loss is lower since the rate of inflation is higher in the present case.

Case 13. Let us now discuss the alternative global strategy: gradualism. This strategy includes the following policy measures: a) The rate of growth of money supply is set to a predetermined value. This is the same measure discussed in case 1, although the value to which m is set is higher ($m=0.025$) and therefore no monetary squeeze is implied. b) The rate of growth of bank advances is set to a predetermined value. This assumption is not at all unrealistic in the Italian case where the degree of control which the monetary authorities may activate on the banking sector is rather high. c) The wage formation mechanism is modified as in case 4, i.e. the rate of growth of wages is predetermined at a value which is progressively lowered and no bonus is handed over to wage earners at the end of each year. d) The marginal tax rate is increased by 5 percentage points. e) The value of the ratio of government consumption to output is lowered by 10 percent. f) An endogenous exchange rate equation is introduced and the rate of change in the exchange rate is set so that the exchange rate steadily depreciates towards

Table 24 - Gradualism

	I	II	III	IV	V
Dlog K	+0.09	-0.06	-0.06	-0.15	-0.20
Dlog Y	-0.12	-0.85	-0.54	-0.33	-1.21
Dlog P	-0.20	+0.71	-0.38	-1.18	-6.17
W/P	-4.40	-9.15	-14.30	-19.50	-24.40
R	+51.80	+68.50	+82.00	+84.40	+180.70
BGS	+37.00	+41.85	+46.10	+50.80	+71.30
PG/T	-8.70	-16.40	-19.80	-22.00	-24.80
NFA	-3.00	-7.70	-11.90	-15.20	-13.00

a 30% higher level (i.e. the same amount of devaluation as in the shock therapy, but at a steady rate), with a mean time-lag of five years.

No strict distinction is now obviously possible between impact and medium term effects. Let us start with the response of the rate of change in prices. On a yearly basis we witness a slight gain (-0.2 percent) on the first year of simulation which is more than offset during the second year. Starting from the third year, however, the inflation gain is increasing and reaches a considerable amount (more than 6 percent on a yearly basis) in the final year. It is worth noting, however, how these outcomes are due to a much more unsteady initial behaviour. As fig. 15 shows, the rate of inflation oscillates rather violently in the first year until these movements even out. Two factors are responsible for this result. The first one is the adoption of a "monetarist" rule of fixed monetary expansion, which, as we have seen in case 1, eliminates the buffer represented by the monetary authorities' reaction to the behaviour of inflation. The second is the higher rate of

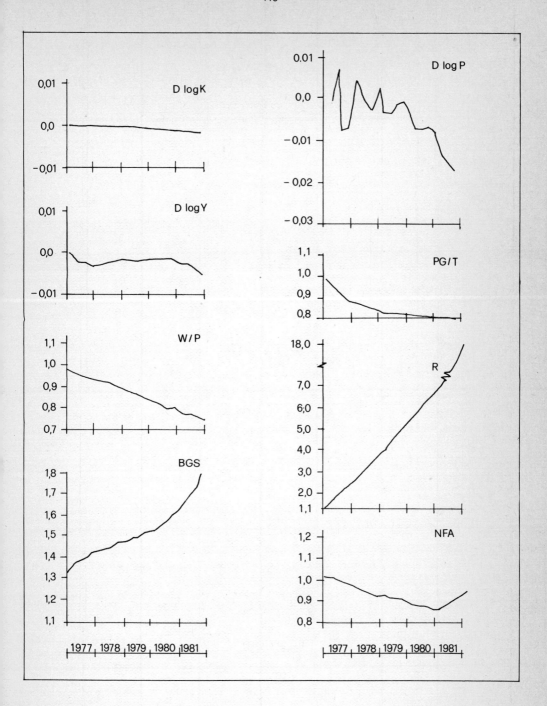

FIG. 15 - Gradualism

inflation with respect to control, due to the steady devaluation of the exchange rate. This second element accounts for the higher amplitude in the oscillations of $D\log P$ with respect to case 1. The increasing gain in inflation is the result of another effect which more than offsets the medium term inflationary bias produced by devaluation. This effect is the predetermination of the rate of increase in wages, which, as we have seen in case 4, is rather effective in curbing inflation. This simulation exercise also seems to suggest that a monetarist rule might contribute to curbing inflation, *if* it is associated with other antiinflationary rules. The loss in real wage, which is considerable and on the increase (almost 25 percent in the final year) is, however, there to show how the true trade-off is, in this case, between $D\log P$ and W/P. The predetermination of the rate of growth of money supply isolates the system, contrary to what happened in the previous case, from the increasing monetary squeeze which the other policy measures would have produced if the monetary authorities' reaction function had been untouched. This sterilized behaviour is the result of an expansion of foreign reserves, which is, however, lower in absolute value than the one resulting in case 12 (shock therapy), and of a decrease in the public sector's borrowing requirement, which is increasing and higher than the result achieved with the shock therapy in the three final years of the simulations. Let us now discuss the mechanisms which produce these results. The public finance policy assumed here produces a direct and increased deflationary effect on the economy, mainly through the curb on private consumption which results from lower disposable income, as well as from lower government consumption. The deflationary tendency is slightly offset in the third and fourth years of simulation as a result of better export performance which is a consequence of the relative fall in domestic output and, partially, of improved price competitiveness. The constant growth rate of credit policy however does not produce satisfactory results as the growth rate of capital stock is curbed rather than enhanced. Given the deflationary tendency in real (and hence expected) output however, it is difficult in this case to discriminate between these two deflationary forces on real capital formation. The general deflationary tendency is also responsible for the current and capital account behaviour. *BGS* improves steadily and more markedly in the two final years. This is essentially the result of lower imports which are produced both by lower domestic growth and higher price competitiveness due to devaluation and lower inflation. The level of net foreign assets decreases steadily during the first four years and then recovers somewhat, although it remains below control. The lower capital outflow is the result of three factors: lower growth, lower inflation and (relatively) lower credit expansion.

As we mentioned above, the simulation exercises were not carried out with the aim of suggesting the implementation of a particular policy and, as a consequence, we will not discuss whether a shock therapy or a gradualistic approach should be applied to the Italian economy. These global policies help us, perhaps to an even greater extent than the previous simulation exercises, to evaluate the workings of the model, in addition to allowing some brief general considerations.

The justification for the proposal, if not the implementation, of global policy strategies usually lies in the fact that the achievement of the policy targets, i.e. the curbing of inflation and/or the elimination of a structural public deficit, requires two kinds of measures[9]: policy measures which directly affect the target, and policy measures which should take care of the side-effects which inevitably emerge as a consequence of the implementation of the first set of measures ("shock absorbers").

As far as the choice between the two strategies is concerned, the monetarist approach (see, for example, Fellner, 1981) maintains that a gradual containment of aggregate monetary demand can be an effective policy insofar as no wage indexation mechanisms are present in the economy. A traditional Keynesian view (Fischer, 1981) maintains that there are two major objections to a shock therapy: the presence of a wage indexation structure once again and the possibility that a drastic cut in aggregate demand *and* an abrupt fall in the inflation rate

---

[9]For a discussion of these topics see Fellner *et al.* (1981)

could lead to a cumulative debt-deflation process, whose costs would far outweigh its benefits. To overcome these difficulties both authors propose that the cut in aggregate demand should be accompanied by some kind of gradualist wage policy and that monetary authorities (Fischer) should be quick to act as lenders of last resort in the case of shock therapies.

From this point of view our simulation exercises (and not just the last two) confirm that, given a dominant cost push component in the inflationary mechanism, the most effective gradualistic approach, in our case, is the modification (in some way or another) of the wage formation mechanisms. As a matter of fact, any shock therapy (such as for instance the price and wage freeze) which is not accompanied by such a modification is bound to produce perverse results as we have seen when discussing case 7. This confirms what is maintained in all the contributions in Fellner *et al.*, 1981. Institutional mechanisms might prevent any global policy from producing the desired results if they are not modified as necessary.

Let us now come to the problem of global monetary demand management and the issue of debt-deflation. Although our model is not particularly well equipped to cover situations of financial distress, our exercises have confirmed the central role of credit and the monetary variables in general. In order to avoid undesirable deflationary effects we have assigned a central role to credit policy in both cases (shock therapy and gradualism). The initial positive effect on real growth in the shock therapy exercise was, in fact, largely due to the initial credit expansion. Subsequent deflation has shown, however, that the behaviour of money stock (as distinguished from the stock of credit) exerts a powerful direct effect on demand via its effect on consumption (and on credit creation itself). This result reproduces quite accurately an institutional feature of the Italian economy where general credit conditions have a relatively low influence on consumption, which, on the contrary, is rather sensitive to monetary conditions (remember that our definition of M includes sight deposits). A rather interesting result in this respect is the deflationary tendency which gradualism (i.e. predetermined rate of growth) in monetary and credit policy produces on the model. This result could obviously have been avoided if $D\log M$ and $D\log A$ had been given higher values, but this would have led to undesirable behaviour of the main target variables, inflation and government deficit. This result suggests that, contrary to monetarist prescriptions, an active monetary and credit policy is badly needed in a financially complex economy. This is in accordance with the problem of coping with debt deflation risks and with unstable price behaviour which resulted in all cases where m was set to a given value. In other words, the gradualist experiment excludes the option of taking the reaction function out of the monetary authorities' hands and the behaviour equation out of the credit system. But there is another, and perhaps more relevant, point that suggests the extreme difficulty of eliminating "activist" financial policies from our model: the high degree of real and financial openness of the economy.

The practical implementation of a medium term financial gradualism implies a sterilization possibility which seems highly unrealistic. In fact such a policy would involve a much higher amount of regulation and control on the variables of the economy (from the capital account to the exchange rate), which we chose not to simulate just because of their scarce realism.

This last consideration leads us to another and final point in our discussion: the possibility of implementing global adjustment policies given the fact that the high degree of openness in the economy *could* be considered as an understatement for "a high degree of dependence". This becomes clearer if one reflects on the fact that the only really feasible demand management policy, be it monetary or fiscal, is a deflationary one: any expansionary demand management would inevitably run against the balance of payments constraint in its two major components.

One should not, however, jump too hastily to this rather obvious conclusion. Overall policies in a highly open economy can be implemented with some success (i.e. policy authorities do maintain some degree of freedom in their action) if due attention is paid to the overall behaviour of the economy. As Gutowsky, Haertel and Scharrer (1981) have stressed, the efficacy of an overall strategy is greatly modified if it is applied to an open economy rather than to a nearly closed one as in the US. In their discussion of the global policy implemented in West Germany over the last decade, they noticed how a global demand restriction had forced German firms to increase their export sales (in spite of a stronger Deutschemark) and this had partly defeated the efficacy of restrictive measures. The improved balance of payments, on the other hand, produced beneficial effects on inflation thanks to the revaluing currency.

Our simulation exercises confirm this behaviour but they also produce some additional insight into this problem. It is also true in our model that a global restraint improves the current (and capital) account and that a revaluation depresses inflation (see cases 8 and 9). The export-led nature of our model is therefore confirmed. But, as we have seen in the case of a higher world demand (case 11), the exogenous export drive might be, at least partially, depressed if internal supply constraints come into force. This suggests that the above statement about the external constraint on internal expansion should be modified as follows. An expansionary domestic policy should be aimed at providing the best possible *accomodation* to an exogenous demand pull mainly by allowing an adequate capital accumulation. In this respect, tight internal monetary (but expecially credit) policies might lead to perverse results in the medium term. Hence, on the basis of our results, one is tempted to reformulate the general philosophy upon which global strategies rest.

The traditional argument in favour of deflationary policies, be they gradualist or not, runs as follows. Demand deflation leads to lower inflation which, in turn, enhances investment thanks to lower costs and also represents the necessary condition for easier monetary conditions. Wage (incomes) policy alone cannot be considered a reliable antiinflationary policy unless it is accompanied by restrictive demand policies (Fellner, 1981; Gutowsky, Haertel and Scharrer, 1981). Our results show, on the contrary, that, given the institutional characteristics embodied in our model and given the behaviour of foreign variables, inflation targets should be pursued mainly through an active wage policy while demand management, carried out in our case through *activist* monetary and credit policies, should be set in a medium term perspective so as to provide adequate accommodation for foreign demand by sustaining capital accumulation. Furthermore, the first policy will sustain the second since, as we have seen on several occasions, a lower rate of inflation produces a substantial improvement in the balance of payments and hence leaves more room for an activist financial policy.

APPENDIX I

I.1 Derivation of the Steady-state and Comparative Dynamics Results

In general, the deterministic model can be written as a first order differential equation system

$$H[D\log X(t), \log X(t), Z(t), \theta] = 0,$$

where $X(t)$ is the vector of the endogenous variables, $Z(t)$ the vector of the exogenous variables, θ the vector of the parameters, H a vector of linear or non-linear differentiable functions. In order to check whether a steady-state solution exists, the exogenous variables are assumed to follow constant growth paths, that is $Z_i(t) = Z_i^* e^{\lambda_i t}$, $\forall i$, where the rates of growth λ_i may be equal or different, positive, negative or zero. A steady-state is then given by a particular solution of the differential equation system having the form $X_i(t) = X_i^* e^{\rho_i t}$, $\forall i$, where the values of X_i^* and of ρ_i are to be determined. Thus the search for the steady-state is an application of the method of undetermined coefficients (see, for example, Gandolfo, 1980, *passim*). In our model, we assume

$$PMGS = PMGS_\circ e^{\lambda_1 t}, \quad PF = PF_\circ e^{\lambda_2 t}, \quad YF = YF_\circ e^{\lambda_3 t}, \quad PROD = PROD_\circ e^{\lambda_4 t}, \quad Q = Q_\circ, \quad UT_a = UT_p,$$

$$i_f = i_f^\circ.$$

Then if we let

$$C = C^* e^{\rho_1 t}, \quad K = K^* e^{\rho_2 t}, \quad Y = Y^* e^{\rho_3 t}, \quad MGS = MGS^* e^{\rho_4 t}, \quad XGS = XGS^* e^{\rho_5 t}, \quad Y = Y^* e^{\rho_6 t},$$

$$P = P^* e^{\rho_7 t}, \quad PXGS = PXGS^* e^{\rho_8 t}, \quad W = W^* e^{\rho_9 t}, \quad i_{TIT} = i_{TIT}^*, \quad A = A^* e^{\rho_{10} t}, \quad NFA = NFA^* e^{\rho_{11} t},$$

$$m = m^*, \quad T = T^* e^{\rho_{12} t}, \quad G = G^* e^{\rho_{13} t}, \quad V = V^* e^{\rho_{14} t}, \quad R = R^* e^{\rho_{15} t}, \quad M = M^* e^{\rho_{16} t}, \quad k = k^*,$$

$$a = a^*, \quad H = H^* e^{\rho_{17} t}, \quad h = h^*, \quad r = r^*,$$

where the ρ's and the starred initial values are undetermined coefficients, by substituting in system (1)-(23) we have

$$\rho_1 = \alpha_1[\log\gamma_1 + \beta_1\rho_6 + \beta_2(\rho_7-\lambda_1)t + \beta_2(\log P^* - \log PMGS_o) + \log(Y^* e^{\rho_6 t} - \frac{T^*}{P^*}e^{(\rho_{12}-\rho_7)t})$$

$$-\rho_1 t - \log C^*] - \alpha_2[\log\gamma_2 - \beta_3\log i^*_{TIT} + \beta_4\log P^* + \beta_4\rho_7 t + \beta_5\log Y^* + \beta_5\rho_6 t - \log M^*$$

$$-\rho_{16} t], \tag{I.1}$$

$$0 = \alpha_3[\alpha'\log\gamma_3 + \alpha'\log\tilde{Y}^* + \alpha'\rho_3 t - \alpha'\log K^* - \alpha'\rho_2 t - k^*], \tag{I.2}$$

$$\rho_3 = \eta[\log Y^* - \log\tilde{Y}^* + (\rho_6 - \rho_3)t], \tag{I.3}$$

$$\rho_4 = \alpha_5[\log\gamma_4 + \beta_6\log P^* - \beta_7\log PMGS_o + \beta_8\log Y^* - \log MGS^* + (\beta_6\rho_7 - \beta_7\lambda_1 + \beta_8\rho_6$$

$$-\rho_4)t] + \alpha_6[\log\gamma_5 + \log\tilde{Y}^* - \log V^* + (\rho_3 - \rho_{14})t], \tag{I.4}$$

$$\rho_5 = \alpha_7[\log\gamma_6 - \beta_9\log PXGS^* + \beta_9\log PF_o + \beta_{10}\log YF_o - \log XGS^* + (\beta_9\lambda_2 - \beta_9\rho_8 + \beta_{10}\lambda_3$$

$$-\rho_5)t - \beta_{11}[\log\gamma_3 + \log Y^* - \log K^* + (\rho_6 - \rho_2)t], \tag{I.5}$$

$$\rho_6 = \alpha_9[\log\tilde{Y}^* - \log Y^* + (\rho_3 - \rho_6)t] + \alpha_{10}[\log\gamma_5 + \log\tilde{Y}^* - \log V^* + (\rho_3 - \rho_{14})t], \tag{I.6}$$

$$\rho_7 = \alpha_{11}[\log\gamma_7 + \beta_{12}\log PMGS_o + \beta_{13}\log W^* - \beta_{14}\log PROD_o - \log P^* + (\beta_{12}\lambda_1 + \beta_{13}\rho_9$$

$$-\beta_{14}\lambda_4 - \rho_7)t], \tag{I.7}$$

$$\rho_8 = \alpha_{13}[\log\gamma_8 + \beta_{15}\log P^* + \beta_{16}\log PF_o - \log PXGS^* + (\beta_{15}\rho_7 + \beta_{16}\lambda_2 - \rho_8)t], \tag{I.8}$$

$$\rho_9 = \alpha_{14}[\log\gamma_9 + \beta_{17}\log P^* - \log W^* + (\beta_{17}\rho_7 + \lambda_5 - \rho_9)t], \tag{I.9}$$

$$0 = \alpha_{15}[\log \gamma_2 - \beta_3 \log i^*_{TIT} + \beta_4 \log P^* + \beta_5 \log Y^* - \log M^* + (\beta_4 \rho_7 + \beta_5 \rho_6 - \rho_{16})t]$$

$$+ \alpha_{16}(\log \gamma_{10} + \log i^o_f - \log i^*_{TIT}), \qquad (I.10)$$

$$\rho_{10} = \alpha_{17}[\log \gamma_{11} + \beta_{18} \log i^*_{TIT} + \log M^* + (\rho_{16} - \rho_{10})t - \log A^*], \qquad (I.11)$$

$$\rho_{11} = \alpha_{18}[\log \gamma_{12} - \beta_{19} \log i^*_{TIT} + \beta_{20} \log i^o_f + \beta_{21} \log P^* + \beta_{21} \log Y^* - \beta_{22} \log PF_o$$

$$- \beta_{22} \log YF_o + \beta_{23} \log Q_o + (\beta_{21}\rho_7 + \beta_{21}\rho_6 - \beta_{22}\lambda_2 - \beta_{22}\lambda_3 - \rho_{11})t - \log NFA^*], \qquad (I.12)$$

$$0 = \alpha_{20}[\delta_1 \log R^* - \delta_1 \log \gamma_{13} - \delta_1 \log PMGS_o - \delta_1 \log MGS^* + \delta_1(\rho_{15} - \lambda_1 - \rho_4)t$$

$$+ \delta_2(\lambda_2 - \rho_7) - m^*], \qquad (I.13)$$

$$\rho_{12} = \alpha_{21}[\log \gamma_{14} + \beta_{24} \log P^* + \beta_{24} \log Y^* - \log T^* + (\beta_{24}\rho_7 + \beta_{24}\rho_6 - \rho_{12})t], \qquad (I.14)$$

$$\rho_{13} = \alpha_{22}[\log \gamma_{15} + \log Y^* - \log G^* + (\rho_6 - \rho_{13})t], \qquad (I.15)$$

$$\rho_{14} V^* e^{\rho_{14} t} = Y^* e^{\rho_6 t} + MGS^* e^{\rho_4 t} - C^* e^{\rho_1 t} - \rho_2 K^* e^{\rho_2 t} - XGS^* e^{\rho_5 t} - G^* e^{\rho_{13} t}, \qquad (I.16)$$

$$\rho_{15} R^* e^{\rho_{15} t} = PXGS^* XGS^* e^{(\rho_5 + \rho_8)t} - PMGS_o MGS^* e^{(\lambda_1 + \rho_4)t} - \rho_{11} NFA^* e^{\rho_{11} t}, \qquad (I.17)$$

$$\rho_2 = k^*, \qquad (I.18)$$

$$a^* = \rho_{10}, \qquad (I.19)$$

$$m^* = \rho_{16}, \qquad (I.20)$$

$$\rho_{17} H^* e^{\rho_{17} t} = P^* G^* e^{(\rho_7 + \rho_{13})t} - T^* e^{\rho_{12} t}, \qquad (I.21)$$

$$h^* = \rho_{17}, \tag{I.22}$$

$$r^* = \rho_{15}. \tag{I.23}$$

System (I.1)-(I.23) will be identically satisfied if, and only if, the two following sets of equations are satisfied[1]:

Set 1

$$\alpha_1[\beta_2(\rho_7-\lambda_1)+(\rho_6-\rho_1)]+\alpha_2(\rho_{16}-\beta_4\rho_7-\beta_5\rho_6)=0, \tag{I.1'}$$

$$\rho_{12}-\rho_7-\rho_6=0, \tag{I.1'bis}$$

$$\rho_3-\rho_2=0, \tag{I.2'}$$

$$\rho_6-\rho_3=0, \tag{I.3'}$$

$$\alpha_5(\beta_6\beta_7-\beta_7\lambda_1+\beta_8\rho_6-\rho_4)+\alpha_6(\rho_3-\rho_{14})=0, \tag{I.4'}$$

$$\beta_9\lambda_2-\beta_9\rho_8+\beta_{10}\lambda_3-\rho_5-\beta_{11}(\rho_2-\rho_6)=0, \tag{I.5'}$$

$$\alpha_9(\rho_3-\rho_6)+\alpha_{10}(\rho_3-\rho_{14})=0, \tag{I.6'}$$

$$\beta_{12}\lambda_1+\beta_{13}\rho_9-\beta_{14}\lambda_4-\rho_7=0, \tag{I.7'}$$

$$\beta_{15}\rho_7+\beta_{16}\lambda_2-\rho_8=0, \tag{I.8'}$$

$$\beta_{17}\rho_7+\lambda_5-\rho_9=0, \tag{I.9'}$$

$$\beta_4\rho_7+\beta_5\beta_6-\rho_{16}=0, \tag{I.10'}$$

$$\beta_{16}-\rho_{10}=0, \tag{I.11'}$$

$$\beta_{21}\rho_7+\beta_{21}\rho_6-\beta_{22}\lambda_2-\beta_{22}\lambda_3-\rho_{11}=0, \tag{I.12'}$$

$$\rho_{15}-\lambda_1-\rho_4=0, \tag{I.13'}$$

$$\beta_{24}(\rho_7+\rho_6)-\rho_{12}=0, \tag{I.14'}$$

---

[1] Eqs. (I.1') and (I.1'bis) both follow from eq. (I.1) if we observe that the disposable income term can be written as

$$\log[e^{\rho_6 t}(Y^* - \frac{T^*}{P^*}e^{(\rho_{12}-\rho_7-\rho_6)t})].$$

$$\rho_6-\rho_{13}=0, \tag{I.15'}$$

$$\rho_1=\rho_2=\rho_4=\rho_5=\rho_6=\rho_{13}=\rho_{14}, \tag{I.16'}$$

$$\rho_{15}=\rho_5+\rho_8=\lambda_1+\rho_4=\rho_{11}, \tag{I.17'}$$

$$\rho_{17}=\rho_7+\rho_{13}=\rho_{12}. \tag{I.21'}$$

Set 2

$$\log C^* = (\beta_2 - \frac{\alpha_2}{\alpha_1}\beta_4)\log P^* + \log(Y^* - \frac{T^*}{P^*}) + \frac{\alpha_2}{\alpha_1}\beta_3 \log i^*_{TIT} - \frac{\alpha_2}{\alpha_1}\beta_5 \log Y^*$$

$$+ \frac{\alpha_2}{\alpha_1}\log M^* + [-\frac{\rho_1}{\alpha_1} + \log\gamma_1 + \beta_1\rho_6 - \beta_2\log PMGS_o - \frac{\alpha_2}{\alpha_1}\log\gamma_2], \tag{I.1''}$$

$$\log K^* = \log \tilde{Y}^* - \frac{k^*}{\alpha'} + \log\gamma_3, \tag{I.2''}$$

$$\log \tilde{Y}^* = \log Y^* - \rho_3/\eta, \tag{I.3''}$$

$$\log MGS^* = \beta_6 \log P^* + \beta_8 \log Y^* + \frac{\alpha_6}{\alpha_5}\log \tilde{Y}^* - \frac{\alpha_6}{\alpha_5}\log V^* + [-\frac{\rho_4}{\alpha_5} + \log\gamma_4$$

$$-\beta_7\log PMGS_o + \frac{\alpha_6}{\alpha_5}\log\gamma_5], \tag{I.4''}$$

$$\log XGS^* = -\beta_9 \log PXGS^* - \beta_{11}\log Y^* + \beta_{11}\log K^* + [-\frac{\rho_5}{\alpha_7} + \log\gamma_6 + \beta_9\log PF_o$$

$$+ \beta_{10}\log YF_o - \beta_{11}\log\gamma_3], \tag{I.5''}$$

$$\log Y^* = (1+\frac{\alpha_{10}}{\alpha_9})\log \tilde{Y}^* - \frac{\alpha_{10}}{\alpha_9}\log V^* + [-\frac{\rho_6}{\alpha_9} + \frac{\alpha_{10}}{\alpha_9}\log\gamma_5], \tag{I.6''}$$

$$\log P^* = \beta_{13}\log W^* + [-\frac{\rho_7}{\alpha_{11}} + \log\gamma_7 + \beta_{12}\log PMGS_o - \beta_{14}\log PROD_o], \tag{I.7''}$$

$$\log PXGS^* = \beta_{15}\log P^* + [-\frac{\rho_8}{\alpha_{13}} + \log\gamma_8 + \beta_{16}\log PF_o], \tag{I.8''}$$

$$\log W^* = \beta_{17}\log P^* + [-\frac{\rho_9}{\alpha_{14}} + \log\gamma_9], \tag{I.9''}$$

$$\log M^* = -(\beta_3 + \frac{\alpha_{16}}{\alpha_{15}})\log i^*_{TIT} + \beta_4 \log P^* + \beta_5 \log Y^* + [\log \gamma_2 + \frac{\alpha_{16}}{\alpha_{15}} \log \gamma_{10}$$
$$+ \frac{\alpha_{16}}{\alpha_{15}} \log i^o_f], \qquad (I.10'')$$

$$\log A^* = \beta_{18} \log i^*_{TIT} + \log M^* + \log \gamma_{11}, \qquad (I.11'')$$

$$\log i^*_{TIT} = \frac{\beta_{21}}{\beta_{19}} \log P^* + \frac{\beta_{21}}{\beta_{19}} \log Y^* - \frac{1}{\beta_{19}} \log NFA^* + \frac{1}{\beta_{19}} [\log \gamma_{12} + \beta_{20} \log i^o_f$$
$$-\beta_{22} \log PF_o YF_o + \beta_{23} \log Q_o], \qquad (I.12'')$$

$$\delta_1 \log R^* = \delta_1 \log MGS^* + m^* + [\delta_1 \log \gamma_{13} - \delta_2(\lambda_2 - \rho_7) + \delta_1 \log PMGS_o], \qquad (I.13'')$$

$$\log T^* = \log P^* + \log Y^* + [\log \gamma_{14} - \rho_{12}/\alpha_{21}], \qquad (I.14'')$$

$$\log G^* = \log Y^* + [\log \gamma_{15} - \rho_{13}/\alpha_{22}], \qquad (I.15'')$$

$$V^* = \frac{1}{\rho_{14}}(Y^* + MGS^* - C^* - \rho_2 K^* - XGS^* - G^*), \qquad (I.16'')$$

$$NFA^* = -R^* + \frac{1}{\rho_{11}} PXGS^* XGS^* - \frac{1}{\rho_{11}} PMGS_o MGS^*, \qquad (I.17'')$$

$$k^* = \rho_2, \qquad (I.18'')$$

$$a^* = \rho_{10}, \qquad (I.19'')$$

$$m^* = \rho_{16}, \qquad (I.20'')$$

$$\rho_{17} H^* = P^* G^* - T^*, \qquad (I.21'')$$

$$h^* = \rho_{17}, \qquad (I.22'')$$

$r^* = \rho_{15}.$ (I.23")

Since the only unknowns present in set 1 are the growth rates ρ_i, this set should be solved first. The values of the ρ_i are then substituted in set 2, which can be solved for the steady-state initial levels.

As regards set 1, its solution is fairly simple and is sketched here; the results are given in table 3 in the text.

From eqs. (I.16'), (I.3'), (I.2'), (I.1'bis) it follows that all the real variables (including expected output and real taxes) must grow at the same rate (eqs. (I.6') and (I.15') are then identically satisfied). Then eq. (I.17') gives $\rho_8 = \lambda_1$; by substituting these results in eq. (I.5') we find the common rate of growth of the real variables. Given eq. (I.10'), and being $\rho_6 = \rho_1$, eq. (I.1') yields $\rho_7 = \lambda_1$. Then eq. (I.10') determines the growth rate of M, which equals that of A by eq. (I.11'). Eq. (I.17') states that international reserves and NFA must grow at the same rate, which equals the sum of the growth rate of prices and the growth rate of the real variables; this is also the growth rate of H and of T by (I.21') and (I.1'bis). Therefore M and A on the one hand, and R, NFA, H, T on the other, may grow at different rates; if we wish all the financial variables to grow at the same rate we must impose the additional constraint $\lambda_1 + [\beta_9(\lambda_2 - \lambda_1) + \beta_{10}\lambda_3] = \beta_4\lambda_1 + \beta_5[\beta_9(\lambda_2 - \lambda_1) + \beta_{10}\lambda_3]$, but this is not necessary for the existence of the steady state. Finally, eq. (I.9') determines the growth rate of W. The remaining eqs. give constraints; these are

$(\beta_6 - \beta_7)\lambda_1 + (\beta_8 - 1)[\beta_9(\lambda_2 - \lambda_1) + \beta_{10}\lambda_3] = 0,$ (i)

$(\beta_{12} - 1)\lambda_1 + \beta_{13}(\beta_{17}\lambda_1 + \lambda_5) - \beta_{14}\lambda_4 = 0,$ (ii)

$(\beta_{15} - 1)\lambda_1 + \beta_{16}\lambda_2 = 0,$ (iii)

$(\beta_{21} - 1)[\lambda_1 + \beta_9(\lambda_2 - \lambda_1) + \beta_{10}\lambda_3] - \beta_{22}(\lambda_2 + \lambda_3) = 0,$ (iv)

$\beta_{24} = 1,$ (v)

Set 2 is non-linear, for it contains both the logs and the natural values of the unknowns but it can be treated analytically by a reduction process. This process starts from the subset formed by eqs. (I.7") and (I.9"), and arrives, after lengthy and tedious manipulations, at an equation in implicit form in which the only unknown is Y^*. Then it can be shown that a solution exists, but this cannot be given in explicit form.

As regards the *comparative dynamics*, that concerning the steady-state growth rates is immediate and has already been expounded in the text. On the contrary, the comparative dynamics of the steady-state initial levels does not yield any interesting result, for the signs of the partial derivatives of these levels (derived by the solution of set 2) will be ambiguous on *a priori* grounds.

I.2 Linearization about the Steady-state

The first step is to write the model in terms of the logarithmic deviations of the original variables from their respective steady-state paths. In fact, a convenient property of our model is that the logarithms of the ratio of the variables to their steady-state paths satisfy a non-linear *autonomous* differential system (so that we can examine its local stability by using a linear approximation: see Gandolfo, 1981, pp. 25ff.). To this purpose we define new variables x_i as follows:

$$x_1 = \log(C/C^* e^{\rho_1 t}),$$

$$x_2 = k - k^*,$$

$$x_3 = \log(\tilde{Y}/\tilde{Y}^* e^{\rho_3 t}),$$

$$x_4 = \log(MGS/MGS^* e^{\rho_4 t}),$$

$$x_5 = \log(XGS/XGS^* e^{\rho_5 t}),$$

$$x_6 = \log(Y/Y^* e^{\rho_6 t}),$$

$$x_7 = \log(P/P^* e^{\rho_7 t}),$$

$$x_8 = \log(PXGS/PXGS^* e^{\rho_8 t}),$$

$$x_9 = \log(W/W^* e^{\rho_9 t}),$$

$$x_{10} = \log(i_{TIT}/i^*_{TIT}),$$

$$x_{11} = \log(A/A^* e^{\rho_{10} t}),$$

$$x_{12} = \log(NFA/NFA^* e^{\rho_{11} t}),$$

$$x_{13} = m - m^*,$$

$$x_{14} = \log(T/T^* e^{\rho_{12} t}),$$

$$x_{15} = \log(G/G^* e^{\rho_{13} t}),$$

$$x_{16} = \log(V/V^* e^{\rho_{14} t}),$$

$$x_{17} = \log(R/R^* e^{\rho_{15} t}),$$

$$x_{18} = \log(K/K^* e^{\rho_2 t}),$$

$$x_{19} = a - a^*,$$

$$x_{20} = \log(M/M^* e^{\rho_{16} t}),$$

$$x_{21} = \log(H/H^* e^{\rho_{17} t}),$$

$$x_{22} = h-h^*,$$

$$x_{23} = r-r^*.$$

It is then possible to transform the original system into an autonomous system in terms of the variables x_i. Consider, for example, eq. (1) of the model

$$D\log C = \alpha_1 \log \gamma_1 e^{\beta_1 D\log Y} \left(\frac{P}{PMGS}\right)^{\beta_2} (Y-T/P) - \alpha_1 \log C + \alpha_2 \log M - \alpha_2 \log \gamma_2 i_{TIT}^{-\beta_3} P^{\beta_4} Y^{\beta_5}.$$

Since, by definition, it is also satisfied by the steady-state paths, we can write

$$D\log C^* e^{\rho_1 t} = \alpha_1 \log \gamma_1 e^{\beta_1 D\log Y^* e^{\rho_6 t}} \left[\frac{P^* e^{\rho_7 t}}{PMGS_o e^{\lambda_1 t}}\right]^{\beta_2} (Y^* e^{\rho_6 t} - T^* e^{\rho_{12} t}/P^* e^{\rho_7 t})$$

$$-\alpha_1 \log C^* e^{\rho_1 t} + \alpha_2 \log M^* e^{\rho_{16} t} - \alpha_2 \log \gamma_2 i_{TIT}^{*-\beta_3} \left[P^* e^{\rho_7 t}\right]^{\beta_4} \left[Y^* e^{\rho_6 t}\right]^{\beta_5}$$

By subtracting this equation from the previous one, noting that the known function of time $PMGS$ is the same in the two equations (so that it cancels out: this is true for all the exogenous variables), and using the x_i variables defined above, we obtain

$$Dx_1 = \alpha_1 [\beta_1 Dx_6 + \beta_2 x_7 + \log \frac{Y-T/P}{Y^* e^{\rho_6 t} - (T^*/P^*)e^{(\rho_{12}-\rho_7)t}} - x_1] + \alpha_2 [-\beta_3 x_{10} + \beta_4 x_7$$

$$+\beta_5 x_6 - x_{20}]. \tag{I.24}$$

In order to linearize the fraction on the r.h.s. we can write it as

$$\log \frac{Y^* e^{\rho_6 t} \frac{Y}{Y^* e^{\rho_6 t}} - T^* e^{\rho_{12} t} \frac{T/T^* e^{\rho_{12} t}}{P^* e^{\rho_7 t} P/P^* e^{\rho_7 t}}}{Y^* e^{\rho_6 t} - (T^*/P^*)e^{(\rho_{12}-\rho_7)t}}. \tag{I.25}$$

If we recall from section I.1 that $\rho_6 = \rho_{12} - \rho_7$, we can further write the expression under consideration as

$$\log \frac{Y^* e^{x_6} - (T^*/P^*) e^{x_{14} - x_7}}{Y^* - (T^*/P^*)} = \log[Y^* e^{x_6} - (T^*/P^*) e^{x_{14} - x_7}] - \log[Y^*$$

$$- (T^*/P^*)]. \qquad (I.26)$$

In order to log-linearize, we consider $Y^* e^{x_6}$ as a single variable, say x, and $(T^*/P^*) e^{x_{14} - x_7}$ as another variable, say y; thus we can apply the formula given in Gandolfo (1981, p. 98) to the first square bracket on the r.h.s. of eq. (I.26) and obtain

$$\log[\ldots] \cong \log[Y^* e^{x_6^o} - (T^*/P^*) e^{x_{14}^o - x_7^o}] + \frac{1}{Y^* e^{x_6^o} - (T^*/P^*) e^{x_{14}^o - x_7^o}} \{Y^* e^{x_6^o} (\log Y^* e^{x_6}$$

$$- \log Y^* e^{x_6^o}) - (T^*/P^*) e^{x_{14}^o - x_7^o} [\log(T^*/P^*) e^{x_{14} - x_7}$$

$$- \log(T^*/P^*) e^{x_{14}^o - x_7^o}]\}. \qquad (I.27)$$

Since $x_6^o = x_7^o = x_{14}^o = 0$ by definition, it follows that

$$\log[\ldots] \cong \log[Y^* - (T^*/P^*)] + \frac{1}{Y^* - (T^*/P^*)} \{Y^* x_6 - (T^*/P^*)[x_{14} - x_7]\}. \qquad (I.28)$$

By substituting (I.28) into (I.26) we obtain

$$\log \frac{\cdots}{\cdots} = \log[\ldots] - \log[Y^* - (T^*/P^*)] \cong \frac{Y^*}{Y^* - (T^*/P^*)} x_6$$

$$- \frac{T^*/P^*}{Y^* - (T^*/P^*)} (x_{14} - x_7). \qquad (I.29)$$

Note, incidentally, that from eq. (I.14") we obtain

$$\log \frac{Y^*}{T^*/P^*} = \rho_{12}/\alpha_{21} - \log Y_{14} = \frac{1}{\alpha_{21}} [\lambda_1 + \beta_9 (\lambda_2 - \lambda_1) + \beta_{10} \lambda_3] - \log Y_{14},$$

and this allows us to compute the coefficients of x_6, x_7, x_{14} without having to compute the solution for the s.s. levels (these, however, are necessary for other linearizations below).

Thus we have, by substituting eq. (I.21) into eq. (I.24),

$$Dx_1 = -\alpha_1 x_1 + \left[\alpha_1 \frac{Y^*}{Y^*(-T^*/P^*)} + \alpha_2 \beta_5\right] x_6 + \alpha_1 \beta_1 Dx_6 + \left[\alpha_1 \beta_2 + \frac{T^*/P^*}{Y^*-(T^*/P^*)} + \alpha_2 \beta_4\right] x_7$$

$$-\alpha_2 \beta_3 x_{10} - \alpha_1 \frac{T^*/P^*}{Y^*-(T^*/P^*)} x_{14} - \alpha_2 x_{20}. \qquad (I.L.1)$$

Consider now the second equation of the model. By the usual procedure,

$$Dx_2 = \alpha_3 \alpha'(x_3 - x_{18}) - \alpha_3 x_2 + \alpha_4 Dx_{19}, \qquad (I.L.2)$$

and similarly for the remaining equations[2]

$$Dx_3 = \eta(x_6 - x_3), \qquad (I.L.3)$$

$$Dx_4 = \alpha_5(\beta_6 x_7 + \beta_8 x_6 - x_4) + \alpha_6(x_3 - x_{16}), \qquad (I.L.4)$$

$$Dx_5 = \alpha_7[-\beta_9 x_8 - x_5 - \beta_{11}(x_6 - x_{18})] + \alpha_8 Dx_{19}, \qquad (I.L.5)$$

$$Dx_6 = \alpha_9(x_3 - x_6) + \alpha_{10}(x_3 - x_{16}), \qquad (I.L.6)$$

$$Dx_7 = \alpha_{11}(\beta_{13} x_9 - x_7) + \alpha_{12} Dx_{13}, \qquad (I.L.7)$$

$$Dx_8 = \alpha_{13}(\beta_{15} x_7 - x_8), \qquad (I.L.8)$$

$$Dx_9 = \alpha_{14}(\beta_{17} x_7 - x_9), \qquad (I.L.9)$$

$$Dx_{10} = \alpha_{15}(-\beta_3 x_{10} + \beta_4 x_7 + \beta_5 x_6 - x_{20}) - \alpha_{16} x_{10}, \qquad (I.L.10)$$

$$Dx_{11} = \alpha_{17}(\beta_{18} x_{10} + x_{20} - x_{11}), \qquad (I.L.11)$$

$$Dx_{12} = \alpha_{18}(-\beta_{19} x_{10} + \beta_{21} x_7 + \beta_{21} x_6 - x_{12}) + \alpha_{19} Dx_{19}, \qquad (I.L.12)$$

$$Dx_{13} = \alpha_{20} \delta_1(x_{17} - x_4) - \alpha_{20} \delta_2 Dx_7 - \alpha_{20} x_{13} + \delta_3 Dx_{22} + \delta_4 Dx_{23}, \qquad (I.L.13)$$

$$Dx_{14} = \alpha_{21}(x_7 + x_6 - x_{14}), \qquad (I.L.14)$$

$$Dx_{15} = \alpha_{22}(x_6 - x_{15}), \qquad (I.L.15)$$

$$Dx_{16} = -(C^*/V^*) x_1 - (K^*/V^*) x_2 + (MGS^*/V^*) x_4 - (XGS^*/V^*) x_5 + (Y^*/V^*) x_6$$
$$-(G^*/V^*) x_{15} - k^*(K^*/V^*) x_{18} + [(C^*/V^*) - (MGS^*/V^*) + (XGS^*/V^*)$$
$$-(Y^*/V^*) + (G^*/V^*) + k^*(K^*/V^*)] x_{16}, \qquad (I.L.16)$$

---

[2]Eqs. (I.L.16), (I.L.17), (I.L.21) require a few manipulations, which are shown below.

$$Dx_{17} = -(PMGS_o MGS^*/R^*)x_4 + (PXGS^*XGS^*/R^*)x_5 + (PXGS^*XGS^*/R^*)x_8$$
$$-(NFA^*/R^*)\rho_{11}x_{12} - (NFA^*/R^*)Dx_{12} + [(PMGS_o MGS^*/R^*) - (PXGS^*XGS^*/R^*)$$
$$+(NFA^*/R^*)\rho_{11}]x_{17}, \qquad\qquad (I.L.17)$$

$$Dx_{18} = x_2, \qquad\qquad (I.L.18)$$

$$x_{19} = Dx_{11}, \qquad\qquad (I.L.19)$$

$$Dx_{20} = x_{13}, \qquad\qquad (I.L.20)$$

$$Dx_{21} = (P^*G^*/H^*)x_7 - (T^*/H^*)x_{14} + (P^*G^*/H^*)x_{15} + [(T^*/H^*)$$
$$-(P^*G^*/H^*)]x_{21}, \qquad\qquad (I.L.21)$$

$$x_{22} = Dx_{21}, \qquad\qquad (I.L.22)$$

$$x_{23} = Dx_{17}. \qquad\qquad (I.L.23)$$

This completes the linearization of the system about the steady-state, which has been used for the stability and sensitivity analyses expounded in section 3.2 of the text; it should be noted that this system can be reduced to a system of 20 first-order differential equations which possesses 20 characteristic roots[3].

We now show the manipulations to obtain eqs. (I.L.16), (I.L.17) and (I.L.21).

As regards eq. (I.L.16), consider eq. (16) of the model and, recalling that $DK = kK$ by definition, write it as

$$\frac{DV}{V} \equiv D\log V = \frac{Y}{V} + \frac{MGS}{V} - \frac{C}{V} - \frac{kK}{V} - \frac{XGS}{V} - \frac{G}{V}.$$

By performing a similar operation on the corresponding steady-state equation and subtracting, we obtain

---

[3] In fact, if we substitute from eqs. (I.L.19) and (I.L.22) into eqs. (I.L.11) and (I.L.21) respectively, these become zero-order equations. If we further substitute from eqs. (I.L.23) and (I.L.12) into eq. (I.L.17) use the fact that — from eq. (I.L.12) — $Dx_{19} = \alpha_{17}(\beta_{18}Dx_{10} - Dx_{20} - Dx_{11})$ and substitute the values of Dx_{10}, Dx_{20}, Dx_{11} from eqs. (I.L.10), (I.L.20) and (I.L.11) respectively, also eq. (I.L.17) becomes a zero-order equation.

$$Dx_{16} = \frac{Y^*e^{\rho_6 t}}{V^*e^{\rho_{14} t}} \left(\frac{Y/Y^*e^{\rho_6 t}}{V/V^*e^{\rho_{14} t}} - 1 \right) + \frac{MGS^*e^{\rho_4 t}}{V^*e^{\rho_{14} t}} \left(\frac{MGS/MGS^*e^{\rho_4 t}}{V/V^*e^{\rho_{14} t}} - 1 \right)$$

$$- \frac{C^*e^{\rho_1 t}}{V^*e^{\rho_{14} t}} \left(\frac{C/C^*e^{\rho_1 t}}{V/V^*e^{\rho_{14} t}} - 1 \right) - \frac{k^*K^*e^{\rho_2 t}}{V^*e^{\rho_{14} t}} \left(\frac{kK/k^*K^*e^{\rho_2 t}}{V/V^*e^{\rho_{14} t}} - 1 \right)$$

$$- \frac{XGS^*e^{\rho_5 t}}{V^*e^{\rho_{14} t}} \left(\frac{XGS/XGS^*e^{\rho_5 t}}{V/V^*e^{\rho_{14} t}} - 1 \right) - \frac{G^*e^{\rho_{13} t}}{V^*e^{\rho_{14} t}} \left(\frac{G/G^*e^{\rho_{13} t}}{V/V^*e^{\rho_{14} t}} - 1 \right). \quad (I.30)$$

Since $k = x_2 + k^*$ by definition, the last but two term in the r.h.s. can be written as

$$\frac{k^*K^*e^{\rho_2 t}}{V^*e^{\rho_{14} t}} \left(\frac{(x_2+k^*)K/k^*K^*e^{\rho_2 t}}{V/V^*e^{\rho_{14} t}} - 1 \right) = \frac{k^*K^*e^{\rho_2 t}}{V^*e^{\rho_{14} t}} \left(\frac{\frac{x_2+k^*}{k^*}e^{x_{13}}}{e^{x_{14}}} - 1 \right)$$

$$= \frac{k^*K^*e^{\rho_2 t}}{V^*e^{\rho_{14} t}} \left(\frac{x_2}{k^*} e^{x_{13}-x_{14}} + e^{x_{13}-x_{14}} - 1 \right).$$

If we remember that $\rho_1 = \rho_2 = \rho_4 = \rho_5 = \rho_6 = \rho_{13} = \rho_{14}$, the following equation is obtained immediately:

$$Dx_{16} = (Y^*/V^*)(e^{x_6-x_{16}}-1) + (MGS^*/V^*)(e^{x_4-x_{16}}-1) - (C^*/V^*)(e^{x_1-x_{16}}-1)$$

$$-(K^*/V^*)x_2 e^{x_{18}-x_{16}} - k^*(K^*/V^*)(e^{x_{18}-x_{16}}-1) - (XGS^*/V^*)(e^{x_5-x_{16}}-1)$$

$$-(G^*/V^*)(e^{x_{15}-x_{16}}-1). \quad (I.31)$$

The application of the usual linearization formulae ($e^{x_i - x_j} = 1 + x_i - x_j$; also remember that any x° is zero by definition) gives, after rearrangement of terms, eq. (I.L.16).

Consider now eq. (17) of the model. If we divide through by R we get $D\log R = PXGS \cdot XGS/R - PMGS \cdot MGS/R + (UT_a - UT_p)/R - DNFA/R$.

By performing a similar operation on the corresponding steady-state equation and subtracting, we obtain

$$Dx_{17} = \left(\frac{PXGS \cdot XGS}{R} - \frac{PXGS*XGS*e^{(\rho_8+\rho_5)t}}{R*e^{\rho_{15}t}} \right) - \left(\frac{PMGS \cdot MGS}{R} - \frac{PMGS \cdot MGS*e^{\rho_4 t}}{R*e^{\rho_{15}t}} \right)$$

$$-\left(\frac{DNFA}{R} - \frac{DNFA*e^{\rho_{11}t}}{R*e^{\rho_{15}t}} \right) = \frac{PXGS*XGS*e^{(\rho_8+\rho_5)t}}{R*e^{\rho_{15}t}} \left(\frac{PXGS \cdot XGS/PXGS*e^{\rho_8 t} \; XGS*e^{\rho_5 t}}{R/R*e^{\rho_{15}t}} \right.$$

$$\left. -1 \right] - \frac{PMGS \cdot MGS*e^{\rho_4 t}}{R*e^{\rho_{15}t}} \left(\frac{PMGS \cdot MGS/PMGS \cdot MGS*e^{\rho_4 t}}{R/R*e^{\rho_{15}t}} - 1 \right)$$

$$-\frac{\rho_{10} NFA*e^{\rho_{11}t}}{R*e^{\rho_{15}t}} \left(\frac{DNFA/DNFA*e^{\rho_{11}t}}{R/R*e^{\rho_{15}t}} - 1 \right). \tag{I.32}$$

Since $\rho_8 + \rho_5 = \rho_{15}$, $\lambda_1 + \rho_4 = \rho_{15}$, $\rho_{11} = \rho_{15}$, and $PXGS/PXGS*e^{\rho_8 t} = e^{x_8}$ etc., we have

$$Dx_{17} = (PXGS*XGS*/R*)(e^{x_8 + x_5 - x_{17}} - 1) - (PMGS_\circ MGS*/R*)(e^{x_4 - x_{17}} - 1)$$

$$-(\rho_{11} NFA*/R*) \left(\frac{\dfrac{DNFA}{NFA} \cdot \dfrac{NFA}{\rho_{11} NFA*e^{\rho_{11}t}}}{R/R*e^{\rho_{15}t}} - 1 \right). \tag{I.33}$$

The third expression in the r.h.s. can be written as

$$-(\rho_{11} NFA*/R*)(1/\rho_{11}) \left(\rho_{11} \frac{\dfrac{NFA}{\rho_{11} NFA*e^{\rho_{11}t}}}{R/R*e^{\rho_{15}t}} D\log NFA - \rho_{11} \right)$$

$$= -(NFA*/R*) \left(\frac{NFA/NFA*e^{\rho_{11}t}}{R/R*e^{\rho_{15}t}} D\log NFA - \rho_{11} \right)$$

$$= -(NFA*/R*)(e^{x_{12} - x_{17}} D\log NFA - \rho_{11})$$

$$= \text{(by adding and subtracting } e^{x_{12} - x_{17}} D\log NFA*e^{\rho_{11}t})$$

$$=-(NFA^*/R^*)[e^{x_{12}-x_{17}}(D\log NFA - D\log NFA^* e^{\rho_{11}t}) + e^{x_{12}-x_{17}}\rho_{11} - \rho_{11}]$$

$$=-(NFA^*/R^*)[e^{x_{12}-x_{17}}Dx_{12} + \rho_{11}(e^{x_{12}-x_{17}}-1)]. \tag{I.34}$$

If we substitute eq. (I.34) into eq. (I.33) we have

$$Dx_{17}=(PXGS^*XGS^*/R^*)(e^{x_8+x_5-x_{17}}-1)-(PMGS_o MGS^*/R^*)(e^{x_4-x_{17}}-1)$$

$$-(NFA^*/R^*)[e^{x_{12}-x_{17}}Dx_{12}+\rho_{11}(e^{x_{12}-x_{17}}-1)]. \tag{I.35}$$

Application of the usual linearization formulae to eq. (I.35) then gives eq. (I.L.17).

Consider finally eq. (21) of the model. If we divide through by we obtain

$$D\log H = \frac{PG}{H} - \frac{T}{H}.$$

By performing a similar operation on the corresponding steady-state equation and subtracting, we obtain

$$Dx_{17} = \frac{P^* e^{\rho_7 t} G^* e^{\rho_{13} t}}{H^* e^{\rho_{17} t}} \left(\frac{PG/P^* e^{\rho_7 t} G^* e^{\rho_{13} t}}{H/H^* e^{\rho_{17} t}} - 1 \right) - \frac{T^* e^{\rho_{12} t}}{H^* e^{\rho_{17} t}} \left(\frac{T/T^* e^{\rho_{12} t}}{H/H^* e^{\rho_{17} t}} \right)$$

$$= \frac{P^* e^{\rho_7 t} G^* e^{\rho_{13} t}}{H^* e^{\rho_{17} t}} (e^{x_7+x_{15}-x_{21}}-1) - \frac{T^* e^{\rho_{12} t}}{H^* e^{\rho_{17} t}} (e^{x_{14}-x_{21}}-1).$$

If we remember that $\rho_7 + \rho_{13} = \rho_{17} = \rho_{12}$, and use the linearization formulae, we have

$$Dx_{17} = \frac{P^* G^*}{H^*}(x_7+x_{15}-x_{21}) - \frac{T^*}{H^*}(x_{14}-x_{21}),$$

whence eq. (I.L.21) follows.

I.3 On the Elimination of $\tilde{Y}$

The presence of the unobservable variable $\tilde{Y}$ gives rise to obvious problems in the estimation procedure. These are usually circumvented (as they were in the previous versions of our model) in the following way: several time-series for $\tilde{Y}$ are generated by solving the differential equation (3) for different *given* values of η; the model is then estimated by using each of these series in turn, and the choice falls on that series (and therefore on the corresponding value of η) which yields the "best" results for the model as a whole.

A more satisfactory procedure, which allows the joint estimation of η together with the other parameters, is the elimination of $\tilde{Y}$ from the model by means of suitable manipulations; these are shown below.

From eq. (6) of the model we have

$$D\log Y = \alpha_9 \log(\tilde{Y}/Y) + \alpha_{10}\log(\frac{\gamma_5 \tilde{Y}}{Y}\frac{Y}{V}) = \alpha_9 \log(\tilde{Y}/Y) + \alpha_{10}\log(\tilde{Y}/Y) + \alpha_{10}\log\gamma_5$$

$$+\alpha_{10}\log(Y/V) = (\alpha_9 + \alpha_{10})\log(\tilde{Y}/Y) + \alpha_{10}\log(Y/V) + \alpha_{10}\log\gamma_5, \qquad (I.36)$$

whence

$$\log(\tilde{Y}/Y) = \frac{1}{\alpha_9 + \alpha_{10}} D\log Y - \frac{\alpha_{10}}{\alpha_9 + \alpha_{10}}\log(Y/V) - \frac{\alpha_{10}\log\gamma_5}{\alpha_9 + \alpha_{10}},$$

which gives

$$\log\tilde{Y} = \frac{1}{\alpha_9 + \alpha_{10}} D\log Y + \frac{\alpha_9}{\alpha_9 + \alpha_{10}}\log Y + \frac{\alpha_{10}}{\alpha_9 + \alpha_{10}}\log V - \frac{\alpha_{10}\log\gamma_5}{\alpha_9 + \alpha_{10}}. \qquad (I.37)$$

Differentiation with respect to time yields

$$D\log\tilde{Y} = \frac{1}{\alpha_9 + \alpha_{10}} D^2\log Y + \frac{\alpha_9}{\alpha_9 + \alpha_{10}} D\log Y + \frac{\alpha_{10}}{\alpha_9 + \alpha_{10}} D\log V. \qquad (I.38)$$

If we equate the r.h.s. of eq. (I.38) to the r.h.s. of eq. (3) of the model and solve for $\log(\tilde{Y}/Y)$, we obtain

$$\log(\tilde{Y}/Y) = -\frac{1}{\eta(\alpha_9 + \alpha_{10})} D^2\log Y - \frac{\alpha_9}{\eta(\alpha_9 + \alpha_{10})} D\log Y - \frac{\alpha_9}{\eta(\alpha_9 + \alpha_{10})} D\log V. \qquad (I.39)$$

By substituting this result into eq. (I.36) and rearranging terms we obtain

$$D^2\log Y = -(\eta+\alpha_9)D\log Y + \alpha_{10}\eta\log Y - \alpha_{10}D\log V - \alpha_{10}\eta\log V + \alpha_{10}\eta\log Y_5. \qquad (I.40)$$

This is the equation to be used in estimation in the place of eq. (16); note that eq. (3) has to be deleted from the model for it has been "used up" in the reduction process (actually, if we substitute eqs. (I.38) and (I.39) in eq. (3), we obtain an identity). The fact that it is a second-order equation can easily be dealt with in the usual way, namely by introducing a new endogenous variable $y=D\log Y$, whence $D^2\log Y = Dy$ etc.

Eq. (I.37) can be used to substitute for $\log Y$ in eqs. (2) and (4) of the model, so that the elimination procedure is complete.

APPENDIX II

II.1 Data Sources and Definitions

II.1.1 Sources of Data

- *AU*   Unpublished, computed by authors
- *BI*   Bollettino (Banca d'Italia)
- *DOT* Direction of Trade (IMF)
- *IFS* International Financial Statistics (IMF)
- *ISCO* Conti economici nazionali (Istituto Nazionale per lo Studio della Congiuntura)
- *ISTAT* Annuario di contabilità nazionale; Bollettino mensile (Istituto Centrale di Statistica)
- *MAIN* Main Economic Indicators (OECD)
- *QNA* Quarterly National Accounts (OECD)

II.1.2 Definition of Series (All Series Are Quarterly)

C = real consumption expenditure (1970 prices) of the private sector. Source: ISTAT.

K = total fixed investment *less* interpolated capital consumption (both at 1970 prices) cumulated on a base stock of 124497 billion (10^9) lire in 1959-IV. Source: fixed investment, ISTAT; capital consumption: authors' interpolation[1] of yearly ISTAT data.

MGS = real imports of goods and services (1970 prices), national accounts basis. Source: ISTAT.

XGS = real exports of goods and services (1970 prices), national accounts basis. Source: ISTAT.

G = real current expenditure (1970 prices) of the public sector. Source: ISTAT.

T = direct taxes *plus* indirect taxes *plus* social security receipts *less* advances for production *less* advances for non production activities all at current prices. Source: authors' interpolation of yearly ISCO data.

---

[1] All interpolations were carried out as explained in Gandolfo (1981, pp. 114ff).

V	= change in real inventories (1970 prices) cumulated on a base stock of 1072.9 billion (10^9) lire in 1959-IV. Source: ISTAT.
Y	= gross domestic product (1970 prices) *less* interpolated capital consumption. Sources: gross domestic product, ISTAT; capital consumption: authors' interpolation of yearly ISTAT data.
P	= domestic product implicit price deflator (1970=100). Source: domestic product at current and 1970 prices, ISTAT.
$PROD$	= value added of industrial sector (1970 prices) divided by industrial sector employment. Source: ISTAT.
$PXGS$	= implicit price deflator of exports of goods and services (1970=100). Source: exports of goods and services at current and 1970 prices, ISTAT.
W	= income from employment at current prices divided by employees in employment. Source: ISTAT.
i_{TIT}	= long term government bonds yield, period average. Source: IFS.
i_f	= US treasury bill rate on new issues, 3 months bill rate. Source: IFS
M	= sum of money *plus* quasi-money. Source: IFS.
A	= credit advances of the banking sector to residents at current prices. Source: BI.
NFA	= net capital outflows at current prices (source: BI) cumulated on a base stock of 15045 billion (10^9) lire in 1959-IV.
E	= period-average market exchange rate, lire per US dollar. Source: IFS.
$PMGS$	= implicit price deflator of imports of goods and services (1970=100). Source: imports of goods and services at current and 1970 prices, ISTAT.
R	= stock of Italy's international reserves, in current lire, computed by cumulating backwards and forwards the overall balance of payments ("movimenti monetari", source: BI) on a base stock of 8615.1 billion (10^9) lire in 1977-II.
Q	= ratio of 3 months forward exchange rate (lire per US dollar) to E. Source: AU.
$(UT_a - UT_p)$	= net unilateral transfers in current lire. Source: BI. We included in this item a correction factor to account for the discrepancy between the national accounts and the balance of payments data on exports and imports of goods and services.
PF	= this index was built, following Houthakker and Magee (1969) and further contribution by Drollas (1976), in order to take account of the competition between rival international

suppliers to each market. Countries considered both as competitors on third markets and as importing economies were 18 OECD countries (i=Italy)

$PF \quad = tc \cdot PF^{\$},$

$$PF^{\$} = \sum_{\substack{j=1 \\ j \neq i}}^{18} w_{ij}^{x} \sum_{\substack{k=1 \\ j \neq k \\ k \neq i}}^{18} w_{jk}^{*} PX_{k}^{\$}, \text{ where}$$

$$w_{jk}^{x} = XGS_{ij} / \sum_{\substack{j=1 \\ j \neq i}}^{18} XGS_{ij},$$

$$w_{jk}^{*} = MGS_{jk} / \sum_{\substack{k=1 \\ k \neq i \\ k \neq j}}^{18} MGS_{jk},$$

$PX_{k}^{\$}$ = export price index of country k. Source: MAIN

XGS_{ij} = Italy's exports of goods and services to country j. Source: DOT.

MGS_{jk} = imports of goods and services of country j from country k. Source: DOT.

tc = trade conversion factor. Source: IFS.

YF = weighted sum of OECD countries' income, i=Italy (see Houthakker and Magee 1969; Drollas, 1976):

$YF = tc \cdot YF^{\$}$

$$YF^{\$} = \sum_{\substack{j=1 \\ j \neq i}}^{18} w_{ij}^{x} Y_{j}^{\$}, \text{ where}$$

tc = defined above,

w_{ij}^{x} = defined above,

$Y_{j}^{\$}$ = real gross national product of country j (1970 prices, $10^{9}\$$). Source: QNA.

H = monetary base created by the government cumulated on a base value of 3819.7 billion lire (10^{9}) in 1959-IV. Source: BI. Note that, given this definition of H, eq. (21) (see table 1) was modified as follows for estimation purposes: $DH = PG-T+CF$ where CF is an exogenous correction factor ($CF \equiv DH-PG+T$). CF takes account both of the fact that $PG-T$ is not the "true" public sector's deficit since PG covers only nominal current expenditure (as public investment is included in DK) and the fact that DH covers only that part of the public deficit that is financed by monetary expansion. This shortcut solution was adopted in order to minimize the number of equations and to avoid problems arising from the non correspondence of national accounts data (from which PG and T are taken) and public finance data (from which H is taken).

t = time trend, where the first quarter of 1960 is -42.5, the following quarter is -41.5 ... the fourth quarter of 1981 is 42.5, so that the sample mean is zero. The sample period covers 1960-I to 1981-IV.

All stocks were measured at the end of period while prices and interest rates are period averages. The logarithms of all variables were deseasonalized according to the procedure developed by Durbin (1963). All series measured at the end of the period were adjusted in order to be consistent with flow data (Gandolfo, 1981, eqs. (30) and (31) of ch. 3). This allows to consider variables which contain both stocks and flows in their definition: for example, the change in the stock of international reserves as the sum of the current account and of the capital account (defined as the change in the stock of net foreign assets) of the balance of payments. The moving average inherent in a model containing flow data was eliminated from data in accordance with the procedure expounded in Gandolfo (1981, eq. (33) of ch. 3). As a consequence the first three observations of each series were eliminated for estimation purposes. The approximate discrete analogue to the continuous model was obtained as expounded in Gandolfo (1981, ch. 3, §3.3.2).

II.2 On the Endogenous Estimation of Productivity

As we mentioned in ch. 2, attempts were made, in earlier stages of estimation, to make $PROD$ an endogenous variable.

In line with our theoretical assumptions we decided to link the rate of growth of productivity to the rate of growth of fixed capital according to Kaldor's (1961) technical progress function *and/or* to the rate of growth of exports following Beckerman (1962). Both approaches are consistent with the medium-term export-led-growth nature of our model. Therefore our endogenous productivity function was

$$\text{Dlog } PROD = f(k, \text{Dlog } XGS),$$

which, for estimation purposes, was written in the log-linear form

$$D\log PROD = \log\gamma_p + \beta'_p k + \beta''_p D\log XGS.$$

Estimation results were not satisfactory: the parameters β'_p and β''_p were never significantly different from zero (even when either one of them was constrained to zero).

A first reason may be the fact that, since we did not wish to include labour in the model, we did not express the stock of fixed capital in *per capita* terms. A second and — in our opinion — more important reason, is that, while *PROD* is defined with respect to the *industrial* sector, K reflects the overall net investment expenditure of the economy and *XGS* are the overall exports of goods and services of the economy (see §II.1). Therefore we believe that our poor estimation results are attributable to data problems rather than to theoretical problems. Further work is necessary, but at this stage of our research we had to leave *PROD* exogenous.

II.3 Simulation Procedures

Case 1 Predetermined rate of growth of money supply.

Eq. (13) was replaced by $Dm = \alpha_{20}(m^* - m)$, where $m^* = 0.015$. This represents a considerable money squeeze since the observed rate of growth of M in the simulation period is 0.026 per quarter.

Case 2 Import prices excluded from wages.

The value of P included in the definition of $\widehat{W}$ in eq. (9) was replaced by a "corrected" price index $P_c = P/PMGS^{\beta_{12}}$.

Case 3 Predetermined rate of increase of wages with bonus.

Eq. (9) was replaced by $D\log W = \lambda_a$, where λ_a is the predetermined rate of increase ($\lambda_a = 0.03$ in the first two years, $\lambda_a = 0.02$ in the following two and $\lambda_a = 0.01$ in the final year). This implies $W(t) = W_0 \exp(\lambda_a t)$. For the first year of simulation W_0 coincides with the value of the control solution at the same point in time. At the end of the year the difference between the actual rate of growth of P and λ_a is used to compute the bonus — if any — to be given to wage earners. The end-of-the-year $W(t)$ plus the bonus constitute the new initial value (W_0) of the wage rate for the following year; and so on. Since the wage rate so computed is the same as that appearing in eq. (7.1), the implicit assumption is that firms pay out the entire bonus.

Case 4 Predetermined rate of increase of wages withouth bonus.

The simulation is identical to that discussed in case 3. No bonus is, however, handed out to wage earners at the end of the year even in the case that λ_a is different from the actual rate of growth of P.

Case 5 Yearly escalator in wage formation.

The adjustment speed α_{14} in eq. (9.1) was set to 0.25 (this implies a mean time-lag of 4 quarters).

Case 6 Constant money wage plus yearly escalator.

Eq. (9) was replaced by $DlogW=0.0$. This implies $W(t)=W(0)$. For the first year of simulation $W(0)$ coincides with the value of the control solution at the same point in time. At the end of the year W is increased so that $(W/P)(t)=(W/P)(0)+b$ where $b=[DlogP \times W(0)]/4$. The first component of $(W/P)(t)$ sets the real wage which results at the end of the year — $(W/P)(t)^*$ — to the level it had at the beginning of the simulation period, the second component is a bonus which takes account (approximately) of the real wage loss occurring in the course of the year if prices increase. The implicit assumption here is that this bonus is gradually handed over to wage earners by firms. The original proposal suggested that interests should be paid to wage earners by firms. We assume that firms set aside the amount to be paid at the end of the year (i.e. the difference $(W/P)(t)-(W/P)(t)^*$) and invest it at the same interest rate to be paid to wage earners (this is consistent with the fact that the model includes only one interest rate). This means that interest payments due to wage earners are not an additional cost to firms (and this is what really matters since wages enter the model as a cost component). This final hypothesis is also equivalent to assuming perfect price foresight by firms.

Case 7 Price and wage freeze.

This simulation includes three different stages. During stage 1 (first two quarters) eq. (7) is replaced by $DlogP=0.0$ and eq. (9) is replaced by $DlogW=0.0$. The assumption is that while domestic price formation can (and should) be controlled, price formation on international markets is left to the choice of exporting firms and so eq. (8) is untouched. During stage 2 (third and fourth quarters) firms are allowed to increase prices only to the extent made necessary by the increase in import prices, so eq. (7) becomes $DlogP = \alpha_{11} \log(\hat{P}/P) + \alpha_{12} Dm$ with $\hat{P} = \gamma_7 PMGS^{\beta_{12}}$ where γ_7 is set equal to one; eq. (9) is $DlogW=0.0$. The assumption made here is that during the first stage of the experiment real wages are constant, therefore profit margins are curtailed by the increase in other costs (i.e. import prices); during the second stage distribution shifts back in favour of profits since wages are kept constant in nominal terms and prices are allowed to rise[2]. In the third stage the model is restored back to its original structure.

Case 8 Freely fluctuating exchange rate.

In order to carry out this simulation the foreign "price" variables (PF, $PMGS$) were expressed as $E \cdot PF_f$ and $E \cdot PMGS_f$ where E is the lira/US dollar spot exchange rate and f denotes the fact that these variables are expressed in dollar terms (see chapter 2, §2.2.5)[3]. Subsequently a new equation — eq. (24) — was introduced in order to simulate an endogenous determination of E. This equation was specified as follows:

$$DlogE = -\alpha_{23} DlogR, \quad \text{where } \alpha_{23} = 90.0.$$

The assumption made here is that the exchange rate reacts to the excess demand for foreign currency — which is proxied by the change in international reserves — i.e. it rises (depreciates) whenever re-

---

[2] The monetary term is retained since we assume that no restriction is imposed on the speed at which firms adjust prices to their (constrained) target value. Therefore no restriction is imposed on the effects which the change of the rate of change of money supply exerts on this speed (see §2.2). The value of γ_7 is set equal to one since we assume that the allowed increase in prices does not include profit margins.

[3] The variable $Q = FR/E$, where FR is the forward rate, was left unchanged since we had to make the assumption that FR behaved so as to make the ratio FR/E unchanged in order to avoid introducing an endogenous determination of FR as well.

serves fall and viceversa.

The value chosen for α_{23} reflects the assumption that the currency market adjusts almost instantaneously. This in turn means that α_{23} must be considered as an adjustment speed (and therefore $1/\alpha_{23}$ is the mean time-lag, equal to one day). In fact, the partial equilibrium level of the exchange rate, say $\hat{E}$, is in this case such that DR (and so $D\log R$)=0. Let then

$DE=\alpha_{23}(\hat{E}-E)$, whence

$D\log E=\alpha_{23}(\frac{\hat{E}}{E}-1)$, and $\hat{E} \gtreqless E$ when $D\log R \lesseqgtr 0$, so that instead of $(\frac{\hat{E}}{E}-1)$ we can use $-D\log R$ as a proxy.

Case 9 Exchange rate reaction function.

The foreign price variables were modified as in the previous case and eq. (24) was modified as follows $D\log E=\delta_5 (D\log PXGS-D\log PF_f)-\delta_6 D\log R$ where $\delta_5=1.0$ and $\delta_6=3.5$. The first term of the equation is based on the assumption that the monetary authorities have a target exchange rate $\hat{E}$ which reflects a sort of purchasing power parity ($\hat{E}=\frac{PXGS}{PF_f}$) given the weight δ_5, i.e. they adjust the exchange rate in order to preserve the competitiveness of the economy. The second term takes account of the fact that in pursuing their target the monetary authorities also have a desired level of reserves and therefore change the exchange rate in order to offset undesired changes in the level of reserves. The value chosen of δ_6 is very close to the estimated value of α_{20} and is based on the assumption that monetary authorities react approximately with the same speed on the money and on the exchange markets.

Case 10 Constant foreign interest rate.

A constant value was given to i_f in the simulation period; this was the value observed in 1976-IV. Recall that the actual value of i_f during the period rose substantially above the chosen value.

Case 11 Higher rate of growth of world demand.

The rate of growth of YF during the simulation period was increased by 3 percentage points per year over the average observed growth rate (which is approximately 6 per cent per year).

Case 12 Shock therapy.

The following set of modifications was introduced: a) the initial value of T was increased by 20 per cent and the initial value of G was decreased so that resulting initial value of $(G-\frac{T}{P})/Y$ was set equal to the average value of the European Community countries; the value of γ_{15} was decreased by 10%; b) an endogenous exchange rate equation (see case 8) was introduced assuming: b1) an initial 30 per cent devaluation with respect to the 1976-IV observed value, b2) fixed exchange rates thereafter (eq. (24) was defined as Dlog$E$=0.0); c) the initial value of A was increased by 20 per cent; d) a price and wage freeze was adopted (see case 7). After four quarters a yearly escalator clause in wage formation was adopted (see case 5); e) the value of γ_{12} was decreased by 90 per cent for the first four quarters.

Case 13 Gradualism.

The following set of modifications was introduced: a) predetermined rate of growth of money supply (m^*=0.025, see case 1); b) predetermined rate of growth of credit. Eq. (11) was replaced by Dlog$A= \lambda_A$ where λ_A=0.030; c) predetermined rate of growth of wages withouth bonus (see case 4); d) the marginal tax rate β_{24} was increased by 5 percentage points; e) the value of γ_{15} was decreased by 10 per cent; f) an endogenous exchange rate equation was introduced, eq. (24), so defined Dlog$E = \alpha_{23} \log \hat{E}/E$ where $\hat{E}$ is the value initially chosen for E in case 12 and α_{23}=0.05. This implies assuming a constantly depreciating exchange rate, steadily increasing towards a 30% higher level with a mean time-lag of five years.

The unsatisfactory results obtained in the estimation of β_9, the relative price elasticity in the export equation have been discussed in chapter 3 where alternative specifications of the equation are also considered. Given the crucial nature of this parameter, however, we carried out a simulation experiment which consisted in replacing the estimated value of β_9 with the value β_9=0.65 (i.e. a value close to that obtained by the authors in previous versions of the model, see Gandolfo and Padoan, 1982a, 1983a) in order to better appreciate the role of the export price-elasticity in our model. Comparing the basic control run with this simulation run we found out that although the model *is* sensitive to a reasonably high value of β_9, its overall behaviour does *not* vary substantially. In particular we observed a more pronounced cyclical behaviour of real exports (and hence of output, which in turn influences the rest of the model), but the dynamic behaviour of the real variables followed the same pattern: the turning points in the levels and in the rates of growth were unaffected in almost all cases and only slightly affected (i.e. they lagged or anticipated by one or two quarters) in the remaining ones.

II.4 Whither Simulations?

Since Lucas' critique (Lucas, 1976) many model builders have felt uneasy when using simulation exercises to help in the evaluation of alternative policies. However, as the purpose of our simulation exercises was to provide insights into the different properties of the model, we have a clear conscience. Nevertheless, as the model may well be used for future policy evaluations, we had a closer look at Lucas' critique which turned out to be less convincing than it was thought to be. As a matter of fact, Lucas' argument has been challenged on several points and the production of simulations has continued at an unfaltering pace.

What follows is by no means a review of the debate on the topic but rather a very brief discussion of some of the most relevant arguments which have been presented in defence of simulation analysis.

Simulation exercises may present different characteristics[4]. In our opinion, these exercises may be classified in four different types: *Type 1* Exogenous variables are assumed to follow a different path from the one actually observed[5]. *Type 2* Endogenous variables other than

---

[4] What follows draws on recent contributions by Mishkin (1979), Sims (1982), Jonson and Trevor (1981). See also Vines, Maciejowsky and Meade (1983), Klein (1983), Hughes Hallet and Rees (1983).

[5] This type could present a further ramification according to whether in-sample or out-of-sample simulations are carried out.

policy variables are used as "forcing variables", i.e. they are forced to follow a given path. *Type 3* Policy changes, which are uncertain and insincere, in the sense that policy makers either do not specify their present and future behaviour or the policy they announce might be different from the one effectively implemented. *Type 4* The policy change is sincere and once and for all; as a consequence the operators have all the time they need to adjust their behaviour to the new policy regime.

The simulation exercises we have presented cover all four types. Cases 10 and 11 fall obviously under type 1, cases 2 to 7 may be classified as type 2 since wages (and prices in case 7) are forced to follow a given path (at least in part). Case 1 is the only "pure" example of a sincere and once-and-for-all policy change and hence falls under type 4. The remaining cases may be classified under type 3 due to the complexity of the policy strategies (cases 12 and 13) and to the fact that although policy authorities usually announce their overall targets (e.g. a lower rate of inflation, a higher rate of growth, etc.), they will seldom announce the exact route they intend to follow for the policy measures involved. This is certainly the case of the managed float simulated in case 9, where it is hardly conceivable that the central bank will let the public know the ways and means of intervention in the currency market. Case 8 (free fluctuation) could be classified as a type 4 exercise, since it assumes a once-and-for-all regime change. It is however a case in which no policy is explicitly announced.

As Mishkin (1979) and Sims (1982) have pointed out, Lucas' critique fully applies only to type 4 simulations[6] since in this case market operators have all the time and the information necessary to adjust their behaviour to the new policy regime[7]. While Mishkin seems to agree

---

[6] None of the mentioned authors however introduces the classification as presented in the text. For further discussion of Sims' argument see Cooley, LeRoy and Raymon (1984).

[7] A related but separate issue is whether the extreme rational expectations assumption which Lucas introduces may be taken seriously or not. We do not wish to enter into the controversy here. For different criticisms of the concept of rational

with Lucas on this point, Sims argues that "permanent shifts in policy regime are by definition rare events" (p. 118) and hence are not in practice relevant. Although regime changes do occur they are seldom believed to be definitive by the public since they depend on political conditions — both at home and abroad — which cannot be considered permanent. Events such as the switch to new monetary rules in the US and Great Britain and cases of changes in the exchange rate regime, e.g. the implementation of the EMS, suggest that regime changes do occur and that they can last long enough for market operators to adapt to them. As a consequence Sims' argument does not really overcome the difficulty of dealing with this type of policy change.

There are, however, other arguments. One is Jonson and Trevor's (1981) point that Lucas' critique is valid only to a certain extent if a structural model is used, as in our case. Another point is that, in the real world, neither the policy maker nor the market operators (and especially the latter) really know how the system as a whole will react to the implementation of a new policy rule (even if this is sincere) and hence it is a useful exercise (if in-sample simulations are carried out) to "rerun history" in order to gather additional information on the behaviour of the system in the spirit of Meade, Maciejowsky and Vines (1983)[8].

Lucas' critique on the contrary does not seem to be applicable when type 3 policy rules are simulated. If the policy maker does not

---

expectations both on methodological and on heuristic grounds see e.g. Buiter (1980), Pesaran (1982), Swamy, Conway, and von zür Muehlen (1984).

[8] Another critique of Lucas' position is advanced by Salmon and Wallis (1980) who argue that since market operators always optimize their behaviour as Lucas assumes, "there will be a one-to-one relation, in the sense of structural constancy, between the utility function and the derived reaction function which the econometrician observes in the aggregate (.....). To suggest that government policy causes structural variations in the agents' *true* reaction function (author's italics) must imply a change in, and hence previous inadequacy of, the utility function of the economic agent. However this effectively denies the existence of such an agent given the premise of constancy" (p. 41 of the original paper).

specify all the details of his planned strategy (and there are several good institutional reasons for not doing so) the market operators will have a much harder time trying to adjust their behaviour. In addition, if the new strategy does not include (or does not only include) regime changes (e.g. a switch from fixed to flexible exchange rates), private operators can never be certain that the new policy will be permanent. We can develop the argument a bit further by recalling Makin's (1976) contribution,who discusses the problems inherent in the definition of the policy maker welfare function . Makin suggests that, in order to define the weight that the policy maker assigns to different targets, a sort of "revealed preference" approach should be used by trying to identify ex-post the numerical values of such weights. What this point suggests is that the policy maker himself quite often does not have a completely clear idea of what his policy should be and, therefore, he will try to define it by a trial-and-error process. If this is the case uncertainty and insincerity might arise not (only) as an explicit choice of the policy maker but from the fact that he needs time to work out his own line of conduct. Therefore even if private operators have enough time to adjust to a given policy this will change simply because the policy maker will react to the system's response. It is purely academic to argue whether or not this game played between market operators and the policy maker will converge (and how soon) to a perfect forsight/perfect knowledge solution.

As far as type 1 simulations are concerned, Mishkin (1979) argues that shocked values of exogenous variables should be consistent with the time series model of historical values if simulation results are to be considered reliable. This is obtained if "innovations" in exogenous variables are adopted rather than once-and-for-all alterations of their path. However, as Jonson and Trevor (1981) argue, this aspect should not be overemphasized if, as it is in our case, the shocks considered produce responses which may be considered within the range of recent experience.

This argument is applicable also when type 2 simulations are carried out, i.e. when some endogenous variables are used as forcing functions.

In addition, one should recall that the "ad hoc" antiinflationary policies considered in cases 2 to 7 assume the *mutual agreement* of both wage earners and entrepreneurs. As a consequence the implementation of such policies should not produce the reactions — apart from free-riding problems — by market operators predicted by Lucas.

REFERENCES

Allen, K. and A. Stevenson, 1974, *An introduction to Italian economy* (Martin Robertson & Co., London).

Andronov, A.A., A.A. Vitt and S.E. Khaikin, 1966, *Theory of oscillators* (Pergamon Press).

Aoki, M., 1976 *Optimal control and system theory in dynamic economic analysis* (North-Holland, Amsterdam), ch. 5.

Athans, M. and P.L. Falb, 1966, *Optimal control* (McGraw-Hill, New York), ch. 9.

Baffi, R.E., 1971, Ways and programmes of monetary action in Italy: A glance at two decades, in: *Verstehen und gestalten der Wirtschaft* (Mohr, Tubingen), 237-53.

Bailey, R.E., V.B. Hall and P.C.B. Phillips, 1980, A model of output, employment, capital formation and inflation, Cowles Foundation discussion paper No. 552.

Banca d'Italia, anni vari, *Relazione annuale* (Banca d'Italia, Roma).

Bank of Italy, various years, *Annual report* (abridged English version).

Beckerman, W., 1962, Projecting Europe's growth, *Economic Journal* 72, 912-25.

Bergstrom, A.R., ed., 1976, *Statistical inference in continuous time economic models* (North-Holland, Amsterdam).

Bergstrom, A.R. and C.R. Wymer 1976, A model of disequilibrium neo-classical growth and its application to the United Kingdom, in: A.R. Bergstrom, ed., 1976, *op. cit.*, 267-327.

Bergstrom, A.R., 1982, Monetary, fiscal and exchange rate policy in a continuous time econometric model of the United Kingdom, paper presented at the International Seminar on "Recent developments in macroeconometric modelling", organised by CGP and CEPREMAP (Paris, 13-15 September 1982). Published in P. Malgrange and P.-A. Muet, eds., 1984, *Contemporary macroeconomic modelling* (Blackwell, Oxford), 183-206.

Bergstrom, A.R., 1983, Gaussian estimation of structural parameters in higher order continuous time dynamic models, *Econometrica* 51, 117-52.

Bergstrom, A.R., 1984, Continuous time stochastic models and issues of aggregation over time, forthcoming in Griliches, Z. and M.D. Intriligator (eds.), *Handbook of Econometrics* (Amsterdam, North-Holland), Vol. 2, ch. 20.

Blundell-Wignall, A., 1984, Exchange rate modelling and the role of asset supplies: The case of the Deutschemark effective rate 1973 to 1981, *Manchester School* 52, 14-27.

Black, S.W., 1983, The use of monetary policy for internal and external balance in ten industrialized countries, in J.A. Frenkel, ed., *Exchange rates and international macroeconomics* (University of Chicago Press for NBER), 189-225.

Buiter, W.H., 1980, The macroeconomics of Dr. Pangloss: A critical survey of the new classical macroeconomics, *Economic Journal* 90, 34-50.

Caranza, C. and A. Fazio, 1983, Methods of monetary control in Italy, 1974-83, in D.R. Hodgman (ed.), *The political economy of monetary policy: national and international aspects* (Federal Reserve Bank of Boston, Conference Series No. 26), 65-88.

Caranza, C., S. Micossi and M. Villani, 1983, La domanda di moneta in Italia: 1963-1981, in Various Authors, *Ricerche sui modelli per la politica economica* (Banca d'Italia, Roma), Vol. II, 401-73.

Carter, R.A.L. and A.L. Nagar, 1977, Coefficients of correlation for simultaneous equation systems, *Journal of Econometrics* 6, 39-50.

Chiesa, G., 1979, Adjustment of real and financial markets in an open economy: A disequilibrium model of the Italian economy. *Metroeconomica* 31, 167-94.

Chow, G.C., 1981, *Econometric analysis by control methods* (Wiley, New York), ch. 16.

Cooley, T.F., S.F. LeRoy and N. Raymon, 1984, Econometric policy evaluation: A note, *American Economic Review* 74, 467-70.

Corden, M., 1977, *Inflation, exchange rates, and the world economy* (University of Chicago Press).

Davidson, P., 1983, Rational expectations: a fallacious foundation for studying crucial decision-making processes, *Journal of Post Keynesian Economics* 5, 182-198.

De Cecco, M., 1967, *Saggi di politica monetaria* (Giuffrè, Milano).

Deardoff, A.V. and R.M. Stern, 1978a, The terms-of-trade effect on expenditure: some evidence from econometric models, *Journal of International Economics* 8, 409-14.

Deardoff, A.V. and R.M. Stern, 1978b, Modelling the effects of foreign prices on domestic price determination: some econometric evidence and applications for theoretical analysis, *Banca Nazionale del Lavoro Quarterly Review* 31, 333-53.

Deleau, M., J.-P. Laffargue, P. Malgrange, G. de Menil, P.-A. Muet, 1980, *Recherches sur les fondements de la modélisation macroéconomique quantitative*, mimeo, CEPREMAP, Paris.

Deleau, M., P. Malgrange and P.-A. Muet, 1982, A study of short run and long run properties of macroeconometric dynamic models by means of an aggregative core model, paper presented at the international seminar on Recent Developments in Macroeconomic modelling (Paris, September 13-15, 1982). Published in P. Malgrange and P.-A. Muet, eds., 1984, *Contemporary macroeconomic modelling* (Blackwell,

Oxford), 214-46.

Demopoulos, G., G. Katsimbris and S. Miller, 1983, Central bank policy and the financing of government budget deficits: A cross-country comparison, Commission of the European Communities, Economic Papers No. 19.

Drollas, L.P., 1976, The foreign trade sector in disequilibrium: A comparative study of sixteen developed countries, unpublished Ph. D. thesis (London School of Economics).

Durbin, J., 1963, Trend elimination for the purpose of estimating seasonal and periodic components in time series, in: M. Rosenblatt, ed., 1963, *Time series analysis* (Wiley, New York), 3-16.

Fazio, A., 1979, La politica monetaria in Italia dal 1947 al 1978, *Moneta e credito* 32, 269-319 (abridged English version: Monetary policy in Italy from 1970 to 1978, *Kredit und Kapital* 12, 145-80).

Fellner, W. *et al.*, 1981, *Shock therapy or gradualism? A comparative approach to anti-inflation policies* (Group of Thirty, New York, Occasional paper No. 8).

Fellner, W, 1981, Shock therapy or gradualism?, in: W. Fellner *et al.*, *op. cit.*, 9-32.

Fischer, S., 1981, The economics of deflation, in: W. Fellner *et al.*, *op. cit.*, 35-54.

Frenkel, J. and R.M. Levich, 1979, Transaction costs and the efficiency of international capital markets: tranquil versus turbulent periods, in M. Sarnat and G.P. Szegö, eds., 1979, *International finance and trade* (Ballinger, Cambridge, Mass.), Vol. I, 53-81.

Friedman, B.M., 1979, Optimal expectations and the extreme information assumption of 'rational expectations' models, *Journal of Monetary Economics* 5, 23-41.

Galbraith, J.K., 1952, *A theory of price control* (McGraw-Hill, New York).

Gandolfo, G., 1979, The equilibrium exchange rate: theory and empirical evidence, in: M. Sarnat and G.P. Szegö, eds., 1979, *International finance and trade* (Ballinger, Cambridge, Mass.), Vol. I, 99-130.

Gandolfo, G., 1980, *Economic dynamics: methods and models* (North-Holland, Amsterdam).

Gandolfo, G., 1981, *Qualitative analysis and econometric estimation of continuous time dynamic models* (North-Holland, Amsterdam).

Gandolfo, G., 1984a, Recent trends in macroeconometric model building for policy planning, in S.B. Dahiya, ed., *Theoretical foundations of development planning* (forthcoming).

Gandolfo, G., 1984b, Feedback policy rules in a continous time macroeconometric model of the Italian economy, paper presented at the second Viennese Workshop on Economic Applications of Control Theory (Vienna, May 16-18, 1984). Forthcoming in G. Feichtinger (ed.),

Economic applications of control theory (North-Holland, Amsterdam).

Gandolfo, G. and P.C. Padoan, 1980, A macrodynamic general disequilibrium model for the determination of the equilibrium exchange rate, in J. Gutenbaum and M. Niezgodka, eds., 1980, *Applications of systems theory to economics, management, and technology* (Polish Scientific Publishers, Warsaw), 217-40.

Gandolfo, G. and P.C. Padoan, 1981a, Un modello macrodinamico dell'economia italiana: aspetti teorici e risultati empirici, *Note economiche*, 1981, No. 1, 11-61.

Gandolfo, G. and P.C. Padoan, 1981b, A continuous time macrodynamic model of the Italian economy, paper presented at the International Conference of the Society for Economic Dynamics and Control (Lyngby, Denmark, June 22-24, 1981).

Gandolfo, G. and P.C. Padoan, 1981c, Rientro dall'inflazione, tasso di cambio e accumulazione: alcune simulazioni con un modello macrodinamico dell'economia italiana, *Rivista Internazionale di Scienze Sociali* 89, n. 4, 483-507.

Gandolfo, G. and P.C. Padoan, 1982a, Policy simulations with a continuous time macrodynamic model of the Italian economy: A preliminary analysis, *Journal of Economic Dynamics and Control* 4, 205-24.

Gandolfo, G. and P.C. Padoan, 1982b, Inventory cycles in a macrodynamic model of the Italian economy, paper presented at the International Symposium on Inventories (Budapest, August 1982); published in A. Chikán, ed., 1984, *Proc. Second Int. Symp. on Inventories* (Publishing House of the Hungarian Academy of Sciences, Budapest), 145-53.

Gandolfo, G. and P.C. Padoan, 1982c, Demand management and exchange rate policy: the Italian experience - A comment on Tullio, *IMF Staff Papers* 29, 467-74; a Reply by Tullio follows on pp. 475-83.

Gandolfo, G. and P.C. Padoan, 1983a, Cyclical growth in a non-linear macrodynamic model of the Italian economy, paper presented at the workshop on Nonlinear models of fluctuating growth: theory and empirical evidence (Siena, Italy, 24-27 March); published in R.M. Goodwin, M. Krüger, A. Vercelli, eds., 1984, *Nonlinear models of fluctuating growth* (Springer-Verlag, Berlin Heidelberg New York Tokyo), 232-52.

Gandolfo, G. and P.C. Padoan, 1983b, Inflation and economic policy in an open economy: some simulations with a dynamic macroeconometric model, paper presented at the 4th IFAC/IFORS/SEDC Conference on The modelling and control of national economies (Washington, D.C., June 17-19, 1983); published in T. Basar and L.F. Pau, eds., 1984, *Dynamic modelling and control of national economies 1983* (Pergamon Press).

Gandolfo, G. and P.C. Padoan, 1984a, Risultati e problemi dell'uso di un modello continuo per la politica economica, paper presented at the Seminar on "Ricerche quantitative per la politica economica" (SADIBA, Perugia, Italy, February 16-18, 1984); forthcoming in the Proceedings of the Seminar (Bank of Italy, Rome).

Gandolfo, G. and P.C. Padoan, 1984b, A disequilibrium model of real and financial accumulation in an open economy, paper presented at the sixth annual Conference of the Society for Economic Dynamics and Control (Nice, France, June 13-15, 1984).

Goodwin, R.M., 1948, Secular and Cyclical Aspects of the Multiplier and the Accelerator, in: *Income, employment and public policy: Essays in honor of A.H. Hansen* (Norton, New York), 108-32.

Gordon, R.J., 1975, The demand for and the supply of inflation, *Journal of Law and Economics* 18, 807-36.

Gordon, R.J., 1976, Recent developments in the theory of inflation and unemployment, *Journal of Monetary Economics* 2, 185-219.

Granger, C.W.J. and P. Newbold, 1977, *Forecasting economic time series* (Academic Press, New York).

Gutowsky, A., H.-H. Haertel and H.-E. Scharrer, 1981, From shock therapy to gradualism, in: W. Fellner et al., op. cit., 57-69.

Haken, H., 1978, *Synergetics-An introduction* (Springer-Verlag, Berlin Heidelberg New York).

Haken, H., 1983, *Advanced synergetics* (Springer-Verlag, Berlin Heidelberg New York Tokyo).

Harberger, A.C., 1950, Currency depreciation, income, and the balance of payments, *Journal of Political Economy* 58, 47-60.

Harvey, A.C. and J.A. Stock, 1983, The estimation of higher order continuous time autoregressive models, London School of Economics Econometric Programme, Discussion paper No. A.38.

Herring, R.J. and R.C. Marston, 1977, *National monetary policies and international financial markets* (North-Holland, Amsterdam).

Hicks, J.R., 1974, *The crisis in Keynesian economics* (Blackwell, Oxford).

Hodgman, D.R., 1974, *National monetary policies and international monetary cooperation* (Little, Brown & Co., Boston).

Horne, J., 1983, The asset market model of the balance of payments and the exchange rate: A survey of empirical evidence, *Journal of International Money and Finance* 2, 89-100.

Houthakker, H.S. and S.P. Magee, 1969, Income and price elasticities in world trade, *Review of Economics and Statistics* 51, 111-25.

Hughes Hallet, A. and H. Rees, 1983, *Quantitative economic policies and interactive planning: A reconstruction of the theory of economic policy* (Cambridge University Press).

Isard, P., 1978, *Exchange-rate determination: A survey of popular views and recent models*, Princeton Studies in International Finance No. 42 (International Finance Section, Princeton University).

Jonson, P.D., 1976, Money and economic activity in the open economy : The United Kingdom 1880-1970, *Journal of Political Economy* 84, 979-1012.

Jonson, P.D. and R.G. Trevor, 1981, Monetary rules: A preliminary analysis, *Economic Record* 57, 150-67.

Kaldor, N., 1961, Capital accumulation and economic growth, in: F.A. Lutz and D.C. Hague, eds., 1961, *The theory of capital* (Macmillan, London), 177-220.

Kalecki, M., 1933, Outline of a theory of the business cycle, in: M. Kalecki, 1969, *Studies in the theory of business cycles, 1933-39* (Blackwell, Oxford), 3-15, and in: M. Kalecki, 1970, *Selected essays on the dynamics of the capitalist economy, 1933-70* (Cambridge University Press), 1-14.

Keynes, J.M., 1937, Alternative theories of the rate of interest, *Economic Journal* 47, 241-52; reprinted in: *Collected Writings* (Macmillan, London, 1973), Vol. XIV, 201-15.

Kirkpatrick, G., 1983, A prototype national economy for a multi country OECD model, International Institute for Comparative Social Research (Berlin), mimeo.

Klein, L.R., 1980, Economic theoretic restrictions in econometrics, paper presented at the International symposium on Criteria for evaluating the reliability of macro-economic models (Pisa, Italy, December 16-18, 1980); published in G.C. Chow and P. Corsi, eds., 1982, *Evaluating the reliability of macro-economic models* (Wiley, New York).

Klein, L.R., 1983, *The economics of supply and demand* (Blackwell, Oxford).

Knight, M.D., 1976, Output, prices and the floating exchange rate in Canada: a monetary approach, International monetary fund research department DM/77/1, mimeo.

Knight, M.D. and C.R. Wymer, 1978, A macroeconomic model of the United Kingdom, *IMF Staff Papers* **25**, 742-78.

Knight, M.D. and D.J. Mathieson, 1979, Model of an industrial country under fixed and flexible exchange rates, in: J. Martin and A. Smith, eds., 1979, *Trade and payments adjustment under flexible rates* (Macmillan, London), 86-119.

Knight, M.D. and D.J. Mathieson, 1983, Economic change and policy response in Canada under fixed and flexible exchange rates, in: J.S. Bhandari and B.H. Putnam, eds., 1983, *Economic interdependence and flexible exchange rates* (MIT Press, Cambridge, Mass.), 500-29.

Koopmans, T.C., 1950, Models involving a continuous time variable, in: Koopmans, T.C., ed., 1950, *Statistical inference in dynamic economic models*, Cowles Commission for Research in Economics, Monograph 10 (Wiley, New York), 384-89.

Kouri, P.J.K., 1983, Balance of payments and the foreign exchange market: A dynamic partial equilibrium model, in: J.S. Bhandari and B.H. Putnam, eds., 1983, *Economic interdependence and flexible exchange rates* (MIT Press, Cambridge, Mass.), 116-56.

Krenin, M.E. and D. Warner, 1983, Determinants of international trade flows, *Review of Economics and Statistics* 65, 96-104.

Krueger, A.O., 1983, *Exchange-rate determination* (Cambridge University Press).

Laursen, S. and L.A. Metzler, 1950, Flexible exchange rates and the theory of employment, *Review of Economics and Statistics* 32, 281-99; reprinted in: L.A. Metzler, 1973, *Collected papers* (Harvard University Press), 275-307. Pages refer to the original.

Lindbeck, A., ed., 1979, *Inflation and employment in open economies* (North-Holland, Amsterdam).

Lucas, R.E., 1976, Econometric policy evaluation: A critique, *Carnegie-Rochester Series on Public Policy* 1, 19-46.

Makin, J.H., 1976, Constraints on formulation of models for measuring revealed preferences of policy makers, *Kyklos* 29, 709-32.

Malinvaud, E., 1982, Ou en est la theorie macroéconomique?, Commission of the European Communities, Economic Papers, No. 8.

Marschak, J., 1950, Statistical inference in economics: An introduction, in: Koopmans, T.C., ed., 1950, *op. cit.*, 1-50.

Masson, P.R., D.E. Rose and J.G. Selody, 1980, Building a small macromodel for simulation: some issues, technical report 22, Bank of Canada.

Meade, J.E., 1957, *Planning and the price mechanism* (Allen & Unwin, London).

Medio, A., 1983, Synergetics and dynamic economic models, paper presented at the Workshop on "Non-linear models of fluctuating growth: theory and empirical evidence" (Siena, Italy, 24-27 March 1983); published in R.M. Goodwin, M. Krueger, A. Vercelli, eds., 1984, *Non-linear models of fluctuating growth* (Springer-Verlag, Berlin Heidelberg New York Tokyo), 166-91.

Milana, C., 1980, *Petrolio e inflazione*, Ispequaderni No. 18 (Istituto di Studi per la Programmazione Economica, Roma).

Mishkin, F.S., 1979, Simulation methodology in macroeconomics: An innovation technique, *Journal of Political Economy* 87, 816-836.

Mussa, M., 1978, On the inherent stability of rationally adaptive expectations, *Journal of Monetary Economics* 4, 307-13.

Muth, J.F., 1960, Optimal properties of exponentially weighted forecasts, *Journal of the American Statistical Association* 55, 299-306.

Muth, J.F., 1961, Rational expectations and the theory of price movements, *Econometrica* 29, 315-35.

Obstfeld, M., 1982, Aggregate spending and the terms of trade: is there a Laursen and Metzler effect?, *Quarterly Journal of Economics* **97**, 25-70.

OECD, 1973, *Monetary policy in Italy* (OECD, Paris).

Pesaran, M.H., 1982, Expectations formation and macroeconometric modelling, paper presented at the International Seminar on "Recent developments in macroeconomic modelling", organized by CGP and CEPREMAP (Paris, 13-15 September 1982); published in P. Malgrange and P.-A. Muet, eds., 1984, *Contemporary macroeconomic modelling* (Blackwell, Oxford), 27-55.

Petit, M.L., 1981, Modelli di squilibrio e metodi di stima: un'applicazione al mercato delle esportazioni italiane, *Note Economiche*, No.1, 62-82.

Phillips, A.W., 1954, Stabilisation policy in a closed economy, *Economic Journal* **64**, 290-323.

Pissarides, C.A., 1972, A model of British macroeconomic policy, 1955-69, *Manchester School* **40**, 245-59.

Reserve Bank of Australia, 1977, *Conference in applied economic research* (RBA, Sydney).

Richard, D., 1980, International adjustment, exchange rates and growth, paper presented at the world congress of the Econometric Society (Aix-en-Provence, August 28-September 2, 1980).

Rose, D.R., J.G. Selody and P.R. Masson, 1982, Simulated macroeconomic consequences of alternative paths of adjustment of Canadian energy prices to world levels, in K. Clinton, ed., 1982, *Proceedings of the fifth Pacific basin central bank economists' conference* (Bank of Canada), 243-84.

Rustem, B., K. Velupillai and J.H. Westcott, 1978, Respecifying the weighting matrix of a quadratic objective function, *Automatica* **14**, 567-82.

Salmon, M. and K.F. Wallis, 1980, Model validation and forecast comparison: theoretical and practical considerations, paper presented at the international symposium on Criteria for evaluating the reliability of macro-economic models (Pisa, December 16-18); published in: G.C. Chow and P. Corsi, eds., 1982, *Evaluating the reliability of macro-economic models*, (Wiley, New York).

Sargent, T. and N. Wallace, 1973, Rational expectations and the dynamics of hyperinflation, *International Economic Review* **14**, 328-50.

Silverberg, G., 1983, Embodied technical progress in a dynamic economic model: the self-organization paradigm, paper presented at the Workshop on "Nonlinear models of fluctuating growth: theory and empirical evidence" (Siena, Italy, 24-27 March 1983); published in R.M. Goodwin, M. Krüger, A. Vercelli, eds., 1984, *Nonlinear models of fluctuating growth* (Springer-Verlag, Berlin Heidelberg New York Tokyo), 192-208.

Sims, C.A., 1982, Policy analysis with econometric models, *Brookings Papers on Economic Activity*, No. 1, 107-152.

Solnik, B., 1979, International parity conditions and exchange risk, in: Szegö, G.P. and M. Sarnat, eds., 1979, *International finance and trade* (Ballinger, Cambridge, Mass.), Vol. I, 53-81.

Svensson, L.E.O. and A. Razin, 1983, The terms of trade and the current account: the Harberger-Laursen-Metzler effect, *Journal of Political Economy* **91**, 97-125.

Swamy, P.A.V.B., R.K. Conway and P. von zur Muehlen, 1984, The foundations of econometrics, Special Studies Paper No. 182, Federal Reserve Board, Washington.

Sylos Labini, P., 1979, Prices and income distribution in manufacturing industry, *Journal of Post Keynesian Economics* **2**, 3-25.

Taylor, J., 1982, Supply disturbances and monetary rules in an open economy: some empirical results, in K. Clinton, ed., 1982, *Proceedings of the fifth Pacific basin central bank economists' conference* (Bank of Canada), 217-39.

Theil, H., 1964, *Optimal decision rules for government and industry* (North-Holland, Amsterdam), ch. 2.

Tullio, G., 1981, Demand management and exchange rate policy: the Italian experience, *IMF Staff Papers* **28**, 80-117.

Vercelli, A., 1982, Is instability enough to discredit a model?, *Economic Notes*, No. 3, 173-189.

Vines, D., J. Maciejowski and J.E. Meade, 1983, *Demand management* (Allen and Unwin, London).

Wymer, C.R., 1972, Econometric estimation of stochastic differential equation systems, *Econometrica* **40**, 565-77. Reprinted in: Bergstrom, A.R., ed., 1976, 81-95.

Wymer, C.R., 1976, Continuous time models in macro-economics: specification and estimation, paper presented at the SSRC-Ford foundation conference on "Macroeconomic policy and adjustment in open economies" (Ware, England, April 28-May 1, 1976).

Wymer, C.R., 1979, The use of continuous time models in economics, unpublished manuscript.

Wymer, C.R., 1983, Estimation of models with distinct states, unpublished manuscript.

Wymer, C.R., various dates, TRANSF, RESIMUL, CONTINEST, PREDIC, APREDIC computer programs and relative manuals; supplement No. 3 (Oct. 1979) on solution non-linear differential equation systems.

Author Index*

Allen, K., 33n
Andronov, A.A., 40
Aoki, M., 29
Athans, M., 29

Baffi, P., 33n
Bailey, R.E., 6n
Banca d'Italia, 33n, 145
Bank of Italy, 30, 33, 48, 60, 61, 101
Bank of Canada, 1
Beckerman, W., 147
Bergstrom, A.R., 1, 5, 20n
Black, S.W., 30, 60n
Blundell-Wignall, A., 52n
Buiter, W.H., 155n

Caranza, C., 33n, 54n, 55
Carter, R.A.L., 42
Chiesa, G., 11n, 36, 42n
Chow, G., 29
Committee on Policy Optimisation, 1
Conway, R.K., 155n
Cooley, T.F., 154n
Corden, M., 46

Davidson, P., 155n
De Cecco, M., 23
Deardoff, A.V., 19n, 24n
Deleau, M., 1
Demopoulos, G., 60, 61
Drollas, L.P., 56, 57, 145, 146
Durbin, J., 147

Falb, P.L., 29
Fazio, A., 33n, 48, 48n
Fellner, W., 113n, 121, 121n, 122, 124
Fischer, S., 116, 121, 122
Frenkel, J., 28n
Friedman, B.M., 22n

Galbraith, J.K., 96n

Gandolfo, G., 1, 1n, 2, 3n, 4n, 5, 9, 10n, 11, 11n, 22, 36, 42, 42n, 43, 45n, 48n, 50n, 56n, 60, 71, 126, 133, 144n, 147, 152
Goodwin, R.M., 4
Gordon, R., 22n, 47n
Granger, C.W.J., 22n
Gruber, J., 29
Gutowsky, A., 124

Haertel, H.-H., 124
Haken, H., 9, 40
Hall, V.B., 6n
Harberger, A.C., 19
Harvey, A.C., 6n
Herring, R.J., 31n, 47n
Hicks, J.R., 25
Hodgman, D.R., 33n
Horne, J., 9, 35, 50n, 51
Houthakker, H.S., 145, 146
Hughes Hallet, A., 153n

IMF, 1, 144
Isard, P., 35, 106
ISCO, 144
ISTAT, 144, 145

Jonson, P.D., 1, 35, 42n, 153n, 155, 156

Kaldor, N., 25n, 147
Kalecki, M., 20n
Katsimbris, G., 60, 61
Keynes, J.M., 21
Kirkpatrick, G., 1
Klein, L.R., 153n
Knight, M.D., 1, 20n, 22n, 35, 42n, 45n, 50, 52n, 76n
Kouri, P.J.K., 106
Koopmans, T.C., 4
Krasovskiĭ, N.N., 40
Kreinin, M.E., 55, 56
Krueger, A.O., 35

*Numbers followed by n refer to footnotes.

Laursen, S., 19, 54
LeRoy, S.F., 154n
Levich, R.M., 28n
Lucas, R.E., 77n, 153, 154, 154n, 155, 155n, 157

Maciejowski, J., 153n, 155
Magee, S.P., 145, 146
Makin, J.H., 29, 156
Malinvaud, E., 22n
Marschak, J., 4
Marston, R.C., 31n, 47n
Masson, P.R., 1
Mathieson, D.J., 42n, 45n
Meade, J.E., 96n, 153n, 155
Medio, A., 9, 40
Metzler, L.A., 19, 54
Micossi, S., 54n, 55
Milana, C., 46
Miller, S., 60, 61
Mishkin, F.S., 153n, 154, 156
Mussa, M., 22n
Muehlen, P. von zur, 155n
Muth, J.F., 22n

Nagar, A.L., 42
Newbold, P., 22n

Obstfeld, M., 19n
OECD, 33n, 144, 146
Padoan, P.C., 1, 9, 11, 11n, 36, 45n, 48n, 50, 50n, 56n, 60, 152
Pesaran, M.H., 155n
Petit, M.L., 57, 81
Phillips, A.W., 29
Phillips, P.C.B., 6n
Pissarides, C.A., 29

Raymon, N., 154n
Razin, A., 19n
Rees, H., 153n
Reserve Bank of Australia, 1, 35, 42n, 76n
Richard, D., 1
Rose, D.R., 1
Rustem, B., 29

Salmon, M., 155n
Sargent, T., 22n
Scharrer, H.-E., 124
Selody, J.G., 1
Silverberg, G., 9, 40
Sims, C.A., 153n, 154, 154n, 155
Solnik, B., 28n
Stern, R.M., 19n, 24n
Stevenson, A., 33n
Stock, J.A., 6n
Student, 43n
Svensson, L.E.O., 19n
Swamy, P.A.V.B., 155n

Taylor, J., 1
Theil, H., 29
Trevor, R.G., 1, 35, 42n, 50, 153n, 155, 156
Tullio, G., 11n, 36, 42n, 76n

Velupillai, K., 29
Vercelli, A., 40n
Villani, M., 54n, 55
Vines, D., 153n, 155

Wallace, N., 22n
Wallis, K.F., 155n
Warner, D., 55, 56
Westcott, J.H., 29
Wymer, C.R., 1, 5, 6, 10n, 20n, 33n, 35, 42, 50, 52n, 57, 61, 76n

Subject Index*

Accumulation
- of capital, 14, 20, 20n, 21, 33, 44-5, 82, 96, 110, 116, 120, 124, 125
- of financial stocks, 13
- of foreign financial assets, 14
- of international reserves, 14, 110
- of real stocks, 13

Adaptive expectations, 21, 22, 22n, 44
Adjustment
- equations, 3, 6-7
- speeds, 8-9, 43ff, 65, 101
- velocity of (see - speeds)

Aggregate demand, 121-2
Antiinflationary policies, 77, 78ff, 124
Asset market approach, 8-9, 35
- and adjustment speeds, 9, 50, 51
Assignment of instruments to targets, 41

Balance of payments, 14, 31, 31n, 37, 37n, 123-4
- and monetary policy, 31, 31n, 79
Bank advances, 13, 21, 23, 27, 28, 48, 48n, 59, 71, 79, 82, 116, 117
Banking System, 27, 59, 110, 117
Bifurcations, 40, 41, 67, 68, 69, 70, 71
Bonus to wage earners, 84, 90, 93-4, 118, 148ff

Capacity utilization, 21, 23n, 55
- effect on exports, 58, 104, 111-2, 124
Capital
- account, 89, 121
- flows, 27-8, 48, 52, 79, 82, 91, 96, 98, 114, 116
- mobility, 9, 51
Capital accumulation (see Accumulation of capital)
Capital/output ratio, 21, 55
Capital stock, 13, 20, 45, 104, 116, 148
Capital stock adjustment principle, 20-1

Carter-Nagar system test, 42-3
Characteritic roots, 39, 62-3, 138
- and sensitivity analysis, 39-40, 63ff
Comparative dynamics, 38-9, 133
Consumption function, 19, 52, 54
- speed of adjustment, 20, 45
Continuous models, 4ff
Cost of foreign inputs, 24-5
Cost-push, 25-100
Cross equation restrictions, 2-3
Credit, 14, 21, 23, 45-6, 104, 110, 113, 114, 120, 122
- rationing, 27, 30
- induced cycles, 71
Current
- account, 89, 93, 96, 104, 114, 121
Cycles
- credit induced, 71
- inventory determined, 69
Cyclical growth, 4, 63

Damping period, 62
Debt-deflation, 116, 122
Demand
- aggregate (see Aggregate demand)
- for exports (see Export function)
- for imports (see Import function)
- for money (see Money demand)
Deposits, 27
Discrepancy between theoretical and econometric modelling, 1-2
Discrete and continuous models, 4ff
Disequilibrium models, 3
Disposable income, 19, 51, 52, 54, 129
Distribution of income (see Income distribution)

Eigenvalues (see Characteristic roots)
Endogenous
- exchange rate determination, 35, 101ff, 114, 118
- variables, 18, 153

*Number followed by n refer to footnotes

Equilibrium
- of asset markets, 8-9, 50, 51
- of balance of payments, 37
- of flows, 106
Equilibrium exchange rate, 106, 151
Excess demand, 48
Exchange market intervention, 101ff, 151
Exchange rate,
- appreciation (revaluation), 104, 105
-- and inflation, 102, 105
- depreciation (devaluation), 101, 104, 114, 118, 120, 152
-- and inflation, 102
- determination, 9, 35, 51, 101ff, 150-1
-, equilibrium (see Equilibrium exchange rate)
- expectations, 28, 28n, 60
-, fixed, 34-5, 114, 114n, 152, 156
-, flexible, 106ff, 150-1, 156
-, forward, 28, 28n, 60, 145, 150n
- managed float, 34-5, 55-6, 101ff, 150-1
Exogenous variables, 19, 153, 156
Expectations, 13, 14, 21, 28, 60, 142
- adaptive, 21, 22, 22n, 44
- coefficient, 22, 44, 142
-, formation of, 21
- rational, 22n, 154n
Expected output, 21, 22, 23-4, 37, 37n, 45, 45n, 142
Export(s)
- function, 23, 56-7
- led growth, 13, 14, 37, 39, 78, 111, 147
- prices, 23, 25-6, 47, 57, 96, 101, 152-3

Fast motion variables, 9, 49
Feedback policy rules, 29ff
FIML, 2, 42
Financial variables, 3, 13, 14, 26ff, 73, 132
Fiscal policy, 33, 65, 113, 118, 152
Fixed exchange rate (see Exchange rate)
Flexible exchange rate (see Exchange rate)
Flow variables, 7
Forcing variables, 154, 157
Foreign
- assets, 27-8, 38, 48, 59-60

- demand (income), 23, 27, 111, 124, 151
- interest rate, 26, 28, 28n, 108ff, 145, 151
- price variables, 23, 34, 58
Forecasts, 10, 72ff
- ex ante and ex post, 11, 72
- in sample and out of sample, 11, 72
- static and dynamic, 10, 11, 72
Forward exchange rate, 28, 28n, 60

Global policy strategies, 77, 113-25
- gradualist approach, 117-21, 152
- shock therapy, 113-7, 152
Government expenditure, 34, 49, 61, 62, 68, 113, 118, 120, 152
Growth-cyclical path, 4, 63
Growth rate(s)
- of endogenous variables in steady state, 36ff, 126ff
- of exogenous variables, 36, 126
- of money supply, 29ff, 36, 38, 44, 58, 59, 60-1, 79ff, 91, 100, 116, 117, 120

Import(s)
- elasticity, 22, 55-6
- function, 22-3
- price, 22
Imported
- inflation, 81ff
- final goods, 25
Income
- adjustment equation, 23
- distribution, 38, 58n
Index of competitiveness, 23, 101, 145-6
Inflation
- differential, 33
- /growth trade-off, 89, 90, 105
- rate, 24, 77, 98, 101, 104
- /wage trade-off, 81, 100, 105
Instantaneous variables, 7
Integral effect in wage rate predetermination, 86, 88, 89, 91
Interest rate, 14, 20, 26, 27, 28, 28n, 47-8, 51, 54, 110, 145
- differential, 27, 48, 54n
-, foreign (see Foreign)
International reserves, 13, 33, 37, 38, 67, 74, 79, 91, 96, 98, 101ff, 106, 117, 145, 150-1

Inventories, 22, 23, 24, 56, 74
- and cycles, 69
Investment function, 20-1, 44-5
-, and capital stock adjustment, 20-1
-, speed of adjustment, 21, 44

Labour productivity, 25, 25n, 58, 147-8
Latent roots (see Characteristic roots)
Least squares
- ordinary, 3
- three stage, 2
Likelihood ratio test, 42n
Linear approximation
-, about steady-state, 39, 133ff
-, of adjustment speeds, 20, 21, 24, 28
Local stability (see Stability)

Managed flexibility, 34-5, 101ff, 150-1, 154
Markup pricing, 24-5, 46, 58
Mean time-lag, 43
Monetarist rule, 79ff, 118
Monetary
-expansion rate, 29ff, 36, 38, 44, 58, 59, 60-1, 79ff, 91
- policy, 29ff, 48, 79ff, 98, 106, 123, 125
Money
- demand, 20, 26, 54-5
- illusion, 22, 55
- stock (supply), 13, 14, 20, 29ff, 59, 67
--, rate of growth of (see Growth rate of money supply)

Net foreign assets, 27-8, 38, 48, 59-60, 74, 89, 121
Non-linear
-forecasts, 71ff
- model 57, 71
- simulations, 77ff

Observation interval, 7, 8, 9
Order variables, 9, 49ff, 71
Output, determination of, 23-4, 46
Overidentifying restrictions, 42n

Partial
- adjustment equations, 31
- equilibrium, 15
Period
- of observation, 7, 8, 9
- of reference, 8
- of the cycle, 62-3
Point variables (see Instantaneous variables)
Policy
- functions, 29ff, 40, 60-1, 64, 79, 96, 101, 102, 148
- parameters, 40-1, 64ff
- rules, 102, 154ff
- simulations, 29, 77ff
Price
- and wage freeze, 96ff, 113, 114
- elasticity, 70
-- of exports, 23, 56-7, 152-3
-- of imports, 22, 55-6
Prices,
- determination of, 24-5
- speed of adjustment, 24
Productivity of labour (see Labour productivity)
Public
- deficit, 13, 32n, 78, 99, 113, 114, 147
- expenditure (see Government expenditure)
- sector's borrowing requirement, 32, 62n, 74, 147

Qualitative analysis, 36ff, 126ff

Ratchet effect, 19
Rationing, 27, 30
Rational expectations, 22n, 154n
Reaction functions of policy maker (see policy functions)
Representative
- domestic interest rate, 26, 47
- foreign interest rate, 26
- monetary variable, 30
Reserves, international (see International reserves)

Sensitivity analysis, 39-41, 63ff
- and bifurcations, 40, 67ff
Shock therapy (see Global policy strategies)

Simulations, 77ff, 153ff
Slaved variables, 9, 49ff
Slaving principle, 9
Slow motion variables, 9, 49ff
Small country hypothesis, 38n
Speculative behaviour, 22, 28
Speeds of adjustment (see Adjustment speeds)
Stability
- of the model, 39, 63
-, structural, 40, 64
Stabilization policies, 29, 41
Steady-state, 36ff, 126ff
Stock
- of fixed capital (see Capital stock)
- of international reserves (see International reserves)
- of inventories (see Inventories)
- of money (see Money)
- of net foreign assets (see Net foreign assets)
Structural change, 34-5, 61
Structural stability (see Stability)
Synergetics, 8, 40, 49ff
- and adjustment speeds, 9, 49ff

Taxation, 49, 61, 65, 118
Terms of trade, 82
- effect on consumption, 19, 19n, 54
Time-lag (see Mean time-lag)

Undetermined coefficients method, 126

Velocity of adjustment (see Adjustment)

Wage
- and price freeze, 96ff, 113
- and price spiral, 58-9, 59n
- indexation, 25, 81ff, 121
-- modifications in, 77, 81, 90
Wage rate, 25, 88
- predetermination of, 84, 93, 113, 117, 124, 148ff
- real, 77, 100, 105
World
- demand, income, price (see Foreign demand, income, price)

Vol. 157: Optimization and Operations Research. Proceedings 1977. Edited by R. Henn, B. Korte, and W. Oettli. VI, 270 pages. 1978.

Vol. 158: L. J. Cherene, Set Valued Dynamical Systems and Economic Flow. VIII, 83 pages. 1978.

Vol. 159: Some Aspects of the Foundations of General Equilibrium Theory: The Posthumous Papers of Peter J. Kalman. Edited by J. Green. VI, 167 pages. 1978.

Vol. 160: Integer Programming and Related Areas. A Classified Bibliography. Edited by D. Hausmann. XIV, 314 pages. 1978.

Vol. 161: M. J. Beckmann, Rank in Organizations. VIII, 164 pages. 1978.

Vol. 162: Recent Developments in Variable Structure Systems, Economics and Biology. Proceedings 1977. Edited by R. R. Mohler and A. Ruberti. VI, 326 pages. 1978.

Vol. 163: G. Fandel, Optimale Entscheidungen in Organisationen. VI, 143 Seiten. 1979.

Vol. 164: C. L. Hwang and A. S. M. Masud, Multiple Objective Decision Making – Methods and Applications. A State-of-the-Art Survey. XII, 351 pages. 1979.

Vol. 165: A. Maravall, Identification in Dynamic Shock-Error Models. VIII, 158 pages. 1979.

Vol. 166: R. Cuninghame-Green, Minimax Algebra. XI, 258 pages. 1979.

Vol. 167: M. Faber, Introduction to Modern Austrian Capital Theory. X, 196 pages. 1979.

Vol. 168: Convex Analysis and Mathematical Economics. Proceedings 1978. Edited by J. Kriens. V, 136 pages. 1979.

Vol. 169: A. Rapoport et al., Coalition Formation by Sophisticated Players. VII, 170 pages. 1979.

Vol. 170: A. E. Roth, Axiomatic Models of Bargaining. V, 121 pages. 1979.

Vol. 171: G. F. Newell, Approximate Behavior of Tandem Queues. XI, 410 pages. 1979.

Vol. 172: K. Neumann and U. Steinhardt, GERT Networks and the Time-Oriented Evaluation of Projects. 268 pages. 1979.

Vol. 173: S. Erlander, Optimal Spatial Interaction and the Gravity Model. VII, 107 pages. 1980.

Vol. 174: Extremal Methods and Systems Analysis. Edited by A. V. Fiacco and K. O. Kortanek. XI, 545 pages. 1980.

Vol. 175: S. K. Srinivasan and R. Subramanian, Probabilistic Analysis of Redundant Systems. VII, 356 pages. 1980.

Vol. 176: R. Färe, Laws of Diminishing Returns. VIII, 97 pages. 1980.

Vol. 177: Multiple Criteria Decision Making-Theory and Application. Proceedings, 1979. Edited by G. Fandel and T. Gal. XVI, 570 pages. 1980.

Vol. 178: M. N. Bhattacharyya, Comparison of Box-Jenkins and Bonn Monetary Model Prediction Performance. VII, 146 pages. 1980.

Vol. 179: Recent Results in Stochastic Programming. Proceedings, 1979. Edited by P. Kall and A. Prékopa. IX, 237 pages. 1980.

Vol. 180: J. F. Brotchie, J. W. Dickey and R. Sharpe, TOPAZ – General Planning Technique and its Applications at the Regional, Urban, and Facility Planning Levels. VII, 356 pages. 1980.

Vol. 181: H. D. Sherali and C. M. Shetty, Optimization with Disjunctive Constraints. VIII, 156 pages. 1980.

Vol. 182: J. Wolters, Stochastic Dynamic Properties of Linear Econometric Models. VIII, 154 pages. 1980.

Vol. 183: K. Schittkowski, Nonlinear Programming Codes. VIII, 242 pages. 1980.

Vol. 184: R. E. Burkard and U. Derigs, Assignment and Matching Problems: Solution Methods with FORTRAN-Programs. VIII, 148 pages. 1980.

Vol. 185: C. C. von Weizsäcker, Barriers to Entry. VI, 220 pages. 1980.

Vol. 186: Ch.-L. Hwang and K. Yoon, Multiple Attribute Decision Making – Methods and Applications. A State-of-the-Art-Survey. XI, 259 pages. 1981.

Vol. 187: W. Hock, K. Schittkowski, Test Examples for Nonlinear Programming Codes. V. 178 pages. 1981.

Vol. 188: D. Bös, Economic Theory of Public Enterprise. VII, 142 pages. 1981.

Vol. 189: A. P. Lüthi, Messung wirtschaftlicher Ungleichheit. IX, 287 pages. 1981.

Vol. 190: J. N. Morse, Organizations: Multiple Agents with Multiple Criteria. Proceedings, 1980. VI, 509 pages. 1981.

Vol. 191: H. R. Sneessens, Theory and Estimation of Macroeconomic Rationing Models. VII, 138 pages. 1981.

Vol. 192: H. J. Bierens: Robust Methods and Asymptotic Theory in Nonlinear Econometrics. IX, 198 pages. 1981.

Vol. 193: J.K. Sengupta, Optimal Decisions under Uncertainty. VII, 156 pages. 1981.

Vol. 194: R. W. Shephard, Cost and Production Functions. XI, 104 pages. 1981.

Vol. 195: H. W. Ursprung, Die elementare Katastrophentheorie. Eine Darstellung aus der Sicht der Ökonomie. VII, 332 pages. 1982.

Vol. 196: M. Nermuth, Information Structures in Economics. VIII, 236 pages. 1982.

Vol. 197: Integer Programming and Related Areas. A Classified Bibliography. 1978 – 1981. Edited by R. von Randow. XIV, 338 pages. 1982.

Vol. 198: P. Zweifel, Ein ökonomisches Modell des Arztverhaltens. XIX, 392 Seiten. 1982.

Vol. 199: Evaluating Mathematical Programming Techniques. Proceedings, 1981. Edited by J.M. Mulvey. XI, 379 pages. 1982.

Vol. 200: The Resource Sector in an Open Economy. Edited by H. Siebert. IX, 161 pages. 1984.

Vol. 201: P. M. C. de Boer, Price Effects in Input-Output-Relations: A Theoretical and Empirical Study for the Netherlands 1949–1967. X, 140 pages. 1982.

Vol. 202: U. Witt, J. Perske, SMS – A Program Package for Simulation and Gaming of Stochastic Market Processes and Learning Behavior. VII, 266 pages. 1982.

Vol. 203: Compilation of Input-Output Tables. Proceedings, 1981. Edited by J. V. Skolka. VII, 307 pages. 1982.

Vol. 204: K.C. Mosler, Entscheidungsregeln bei Risiko: Multivariate stochastische Dominanz. VII, 172 Seiten. 1982.

Vol. 205: R. Ramanathan, Introduction to the Theory of Economic Growth. IX, 347 pages. 1982.

Vol. 206: M.H. Karwan, V. Lotfi, J. Telgen, and S. Zionts, Redundancy in Mathematical Programming. VII, 286 pages. 1983.

Vol. 207: Y. Fujimori, Modern Analysis of Value Theory. X, 165 pages. 1982.

Vol. 208: Econometric Decision Models. Proceedings, 1981. Edited by J. Gruber. VI, 364 pages. 1983.

Vol. 209: Essays and Surveys on Multiple Criteria Decision Making. Proceedings, 1982. Edited by P. Hansen. VII, 441 pages. 1983.

Vol. 210: Technology, Organization and Economic Structure. Edited by R. Sato and M.J. Beckmann. VIII, 195 pages. 1983.

Vol. 211: P. van den Heuvel, The Stability of a Macroeconomic System with Quantity Constraints. VII, 169 pages. 1983.

Vol. 212: R. Sato and T. Nôno, Invariance Principles and the Structure of Technology. V, 94 pages. 1983.

Vol. 213: Aspiration Levels in Bargaining and Economic Decision Making. Proceedings, 1982. Edited by R. Tietz. VIII, 406 pages. 1983.

Vol. 214: M. Faber, H. Niemes und G. Stephan, Entropie, Umweltschutz und Rohstoffverbrauch. IX, 181 Seiten. 1983.

Vol. 215: Semi-Infinite Programming and Applications. Proceedings, 1981. Edited by A. V. Fiacco and K. O. Kortanek. XI, 322 pages. 1983.

Vol. 216: H. H. Müller, Fiscal Policies in a General Equilibrium Model with Persistent Unemployment. VI, 92 pages. 1983.

Vol. 217: Ch. Grootaert, The Relation Between Final Demand and Income Distribution. XIV, 105 pages. 1983.

Vol. 218: P. van Loon, A Dynamic Theory of the Firm: Production, Finance and Investment. VII, 191 pages. 1983.

Vol. 219: E. van Damme, Refinements of the Nash Equilibrium Concept. VI, 151 pages. 1983.

Vol. 220: M. Aoki, Notes on Economic Time Series Analysis: System Theoretic Perspectives. IX, 249 pages. 1983.

Vol. 221: S. Nakamura, An Inter-Industry Translog Model of Prices and Technical Change for the West German Economy. XIV, 290 pages. 1984.

Vol. 222: P. Meier, Energy Systems Analysis for Developing Countries. VI, 344 pages. 1984.

Vol. 223: W. Trockel, Market Demand. VIII, 205 pages. 1984.

Vol. 224: M. Kiy, Ein disaggregiertes Prognosesystem für die Bundesrepublik Deutschland. XVIII, 276 Seiten. 1984.

Vol. 225: T. R. von Ungern-Sternberg, Zur Analyse von Märkten mit unvollständiger Nachfragerinformation. IX, 125 Seiten. 1984

Vol. 226: Selected Topics in Operations Research and Mathematical Economics. Proceedings, 1983. Edited by G. Hammer and D. Pallaschke. IX, 478 pages. 1984.

Vol. 227: Risk and Capital. Proceedings, 1983. Edited by G. Bamberg and K. Spremann. VII, 306 pages. 1984.

Vol. 228: Nonlinear Models of Fluctuating Growth. Proceedings, 1983. Edited by R. M. Goodwin, M. Krüger and A. Vercelli. XVII, 277 pages. 1984.

Vol. 229: Interactive Decision Analysis. Proceedings, 1983. Edited by M. Grauer and A. P. Wierzbicki. VIII, 269 pages. 1984.

Vol. 230: Macro-Economic Planning with Conflicting Goals. Proceedings, 1982. Edited by M. Despontin, P. Nijkamp and J. Spronk. VI, 297 pages. 1984.

Vol. 231: G. F. Newell, The M/M/∞ Service System with Ranked Servers in Heavy Traffic. XI, 126 pages. 1984.

Vol. 232: L. Bauwens, Bayesian Full Information Analysis of Simultaneous Equation Models Using Integration by Monte Carlo. VI, 114 pages. 1984.

Vol. 233: G. Wagenhals, The World Copper Market. XI, 190 pages. 1984.

Vol. 234: B. C. Eaves, A Course in Triangulations for Solving Equations with Deformations. III, 302 pages. 1984.

Vol. 235: Stochastic Models in Reliability Theory. Proceedings, 1984. Edited by S. Osaki and Y. Hatoyama. VII, 212 pages. 1984.

Vol. 236: G. Gandolfo, P.C. Padoan, A Disequilibrium Model of Real and Financial Accumulation in an Open Economy. VI, 172 pages. 1984.

Vol. 237: Misspecification Analysis. Proceedings, 1983. Edited by T. K. Dijkstra. V, 129 pages. 1984.